MICROSCOPY HANDBOOKS 18

Autoradiography
A Comprehensive Overview

John R. J. Baker

Ciba-Geigy Pharmaceuticals
Wimblehurst Road
Horsham
West Sussex RH12 4AB

Oxford University Press · Royal Microscopical Society · 1989

Oxford University Press, Walton Street, Oxford OX2 6DP

Oxford New York Toronto
Delhi Bombay Calcutta Madras Karachi
Petaling Jaya Singapore Hong Kong Tokyo
Nairobi Dar es Salaam Cape Town
Melbourne Auckland

and associated companies in
Berlin Ibadan

Oxford is a trade mark of Oxford University Press

Published in the United States
by Oxford University Press, New York

Transferred to Digital Printing 2006

British Library Cataloguing in Publication Data
Baker, John R. J.
Autoradiography: a comprehensive overview.
1. Biology. Autoradiography
I. Title II. Royal Microscopical Society
III. Series
574'.028
ISBN 0–19–856422–8

Library of Congress Cataloging in Publication Data
Baker, John R. J.
Autoradiography: a comprehensive overview/John R. J. Baker.
p. cm.—(Microscopy handbooks; 18)
Includes bibliographies and index.
1. Autoradiography. 2. Histochemistry—Technique. I. Royal
Microscopical Society (Great Britain) II. Title. III. Series.
[DNLM: 1. Autoradiography—methods. QU 25 B167a]
QP519.9.A94B35 1989   578—dc20   89-3315
ISBN 0–19–856422–8 (pbk.)

Typeset by Cotswold Typesetting Ltd, Cheltenham

Publisher's Note
The publisher has gone to great lengths to ensure the quality of this reprint
but points out that some imperfections in the original may be apparent

Acknowledgements

I should like to thank the many people who directly or indirectly have helped to make this book possible.

As an industrial researcher my main interest in autoradiography is to apply established methods to increase understanding of the mode of action of pharmaceuticals. During the course of my research I have been fortunate to collaborate with some notable innovators such as Dr Tim Appleton, Dr Nick Blackett, and Professor Mike Williams to whom I extend my sincere gratitude for stimulating my interest in the subject.

No biologist practising autoradiography can be an 'island'. I have been particularly fortunate in working in a research centre alongside skilled radiochemists and I must thank Dr Roy Wade and Dr Derek Brundish for providing me with radiochemicals of high activity and purity over a period of many years.

Finally I thank the many colleagues who have given me enthusiastic technical support, especially Mr Robin Christian, and also my secretary, Miss Emma Mote, for assistance with preparation of the manuscript.

This book is dedicated to my family: to my wife, Gabriella, for her forbearance concerning all the things I might have done instead and to my sons Stuart and Andrew for using their computer as a word processor.

Contents

1 Introduction

Autoradiography is the localization within a solid specimen of a radiolabel by placing the specimen against a layer of detector material (Fig. 1).

Although autoradiographs are usually visualized with the aid of some form of microscope, this is not invariably so. Indeed, rather than being a single technique, autoradiography is a collection of techniques with certain features in common. Different investigators use particular levels of auto-radiography and it will be the purpose of this handbook to describe these to help the beginner choose the correct approach for his particular application.

Although the scope of the book restricts the amount of detail which may be included, an attempt will be made to lay emphasis on certain aspects of the appraisal of autoradiographs. In addition, the committed reader will find extra information in the further reading list at the end of most chapters.

The principle of the autoradiographic process is that radioactive decays taking place within the specimen emit particles of radiation which, after a suitable exposure time, produce a useful number of changes in the detector layer. These alterations within the detector are then amplified in such a way as to make them visible with or without aid to the naked eye. The image produced relates to both distribution and quantity of radiolabel and thus provides information not available from radiometric methods (e.g. scintillation or gamma counting).

Autoradiography has become the almost exclusive preserve of the biologist due to the versatility of tritium and carbon-14 in the radio-synthesis of organic molecules. Moreover, it is convenient that the decay characteristics of these two radionuclides permit their ready detection in the majority of autoradiographic situations.

The characteristics of radiation produced by radionuclides which determine their detectability are mass, energy, and direction. Charge appears to have negligible effect on interaction with the detector.

1.1. Non-photographic detectors

Almost all autoradiographic experiments use a detector composed of some kind of photographic layer. However, in certain applications, notably in health physics monitoring in the nuclear industry, layers of plastic are used. This is possible because fissionable emitters such as uranium-234 produce alpha-particles. These particles (equivalent to helium nuclei) have, in

1

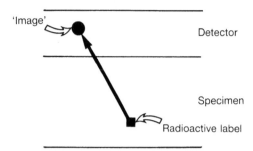

Fig. 1. Generalized schematic diagram showing a vertical section through an autoradiograph

nuclear terms, large mass and high initial energy and can penetrate thin layers of plastics like nitrocellulose, Lexan, and CR-39 producing short straight tracks. The latter are effectively channels which are invisible but may be made visible in the light microscope by etching in a mixture of sodium hydroxide and ethanol. The final result is a series of 'lines' radiating from points of decay, superimposed on a histological section of contaminated tissue.

1.2. Photographic emulsions as detectors

1.2.1. Emulsions

Many salts can be sensitized by light or other electromagnetic wavelengths but none have been exploited as much as the silver halides (AgI, AgBr, AgCl) and nowhere more than in the photographic process. Whereas photographic emulsions are generally crystalline halide mixtures dispersed in gelatin, the nuclear emulsions which we use in autoradiography are relatively pure and contain mainly silver bromide.

Exposure of such an emulsion to light (or other radiation) results in the formation of metallic silver only in those crystals in which energy is deposited by irradiation—the so-called 'latent image'. These latent images are submicroscopic and need to be converted into a visible image by the process of development.

If the halide crystals in the emulsion are perfect, radiation does not produce a latent image. Therefore faults, otherwise known as 'sensitivity specks', are introduced during manufacture into the crystal lattice. The sensitivity specks are commonly composed of silver sulphide and may be on the surface of the crystals or at some depth.

Many of our concepts of the photographic process are at the theoretical level rather than representing true understanding. Thus, current theory states that latent images exist as metallic silver which is produced at the sensitivity specks. After the latent image has been amplified by development the unexposed crystals are removed by dissolution in the fixer.

1.2.2. The role of gelatin in emulsions

Gelatin is a complex protein extracted from cattle hide and bones or pigskin. It is formed from alkaline hydrolysis of the collagen content in the case of bovine tissues or by acid hydrolysis of pigskin collagen.

No matter what the proportion of gelatin in an emulsion the former will always totally encapsulate the silver halide crystals so that each is isolated from its neighbours. The properties of gelatin allow the penetration of chemical agents to the crystals and in a final developed and fixed autoradiograph permit clear visualization of silver grains and morphology of the specimen.

When metallic latent images are formed, it follows that elemental bromine (Br_2) will be given off. The acceptance of the bromine by gelatin reduces the probability that bromine will become re-ionized by electrons and thus recombine to form AgBr.

The fact that gelatin undergoes gel/sol interchange just above room temperature means that it can be melted, diluted, and poured conveniently at temperatures which have little effect on the silver halide crystals. Another important property which gelatin confers upon emulsions is its ability to undergo volume changes when dried or hydrated.

Manufacturers of nuclear emulsions add plasticizers such as glycerol to their products to reduce shrinkage which occurs on drying. Surfactants are also added which are only incompletely washed out at the end of manufacture. This residue of surfactant tends to reduce the rate of drying of emulsions and in so doing helps to minimize the formation of pressure artifacts (see Chapter 4).

2 Components of autoradiographs

2.1. The specimen

Many types of specimen can be studied by autoradiography. For the biochemist, the specimen will often be a chromatogram or electrophoretogram in which radiolabelled metabolites are being separated and identified.

For the biologist, the specimen can be a whole organism or a slice of the whole organism. In the examples so far mentioned a microscope will probably not be necessary to give the investigator the information he wants. In light microscopic (LM) autoradiography, cell smears, 0.5–5 μm sections of fixed embedded material or cryosections can be used. The typical specimen in electron microscopic (EM) autoradiography is the ultra-thin section (40–80 nm) of plastic-embedded material. In rare instances however, ultra-thin cryosections have been used as have freeze-fracture samples (with their replicas) and even bulk specimens in the scanning electron microscope.

The most important of these approaches will be described in the following chapters.

2.1.1. Radiochemicals—choice, storage, and administration

Radiochemicals are generally expensive to buy or make and some thought should go into their selection, handling, and use.

The position of the radiolabel within the molecule must be carefully considered in the light of probable metabolism. Also, if the radiochemical is not carrier-free, an excessive quantity of carrier (unlabelled, but otherwise identical chemical) may prevent adequate uptake of label by the specimen.

After prolonged storage, the purity of radiochemicals should not be taken for granted. The conditions of storage themselves will have an important effect on radiochemical purity. Tritiated compounds which often have a high specific radioactivity, are particularly subject to radiolysis (self-decomposition) which can be reduced by storage in liquid nitrogen.

The use of organic solvents, such as benzene, for the dissolution and storage of, for example, lipids should be viewed with caution. Temperatures below 4°C will crystallize the benzene and concentrate the radiolabel between the crystals thereby increasing the rate of radiolysis. Benzene and other solvents, such as ethanol, should be reduced in concentration or removed so that a minimum of damage is done to the organism receiving the radiolabel.

Routes of administration of radiochemicals should be chosen to be physiologically appropriate. For example, there is little point in injecting radiopharmaceuticals intravenously into animals when the likely human dose form will be oral.

When experimental design allows, much time and cost may be saved by labelling specimens *in vitro*. In this way, high local specific activities can often be achieved in small tissue or cell samples using only a microcurie (37 kBq) of radionuclide.

2.1.2. Is the specimen labelled?

This question appears banal, but it is fundamental and should be carefully considered before leaving specimens to expose for days, weeks, months, or even years.

There may be circumstances in which fixation or some other preparative process can lead to a total or partial loss of radiolabel from the specimen. When in doubt, it is wise to spend a little time using a radiometric method to check that there are sufficient counts within the specimen to make autoradiography feasible. In the case of the so-called 'gamma-emitters' such as iodine-125, sections or bulky samples can be monitored directly in a suitable gamma-counter. On the other hand, if the radionuclide is a soft beta-emitter, e.g. ^3H or ^{14}C, sections of tissue must be dissolved in a cocktail such as Lumagel and counted in a scintillation counter.

2.1.3. Fate of radiochemicals

Whether or not a radionuclide is retained in a biologically relevant location within a specimen depends on metabolism of the parent radiochemical and the method of preparation of the specimen (usually for microscopy). From these considerations two basic types of experiment can be identified:

1. Experiments in which a water-soluble radiolabelled precursor is synthesized into an insoluble macromolecule and unincorporated precursor is eliminated during aqueous preparation of the specimen. An example would be the incorporation of radiolabelled thymidine into DNA.

2. Experiments in which radiolabelled physiological or pharmacological mediators (e.g. drugs, hormones, transmitters, ions) are studied in relation to their distribution or receptor binding affinity.

Experiments of the first type are generally unambiguous in that all decays in the specimen emanate from one radiochemical which is macromolecular with the radionuclide covalently bound and well fixed within the specimen. The second category is a mixture of difficulty and uncertainty. The majority of radiochemicals will be subject to some form of metabolism and will often be washed out of the specimen by aqueous preparation. Thus the origins of a given silver grain in the resulting autoradiograph may be open to serious question. For example, the labelled chemical species may be parent substance or a metabolite and may have been translocated during preparation.

Major strategies for overcoming these problems will be addressed in the coming chapters but it should be apparent that many autoradiographic experiments of the second type need parallel biochemical studies to allow rational interpretation.

2.2. Radionuclides

The choice of radionuclide (radioisotope) for a particular experiment will depend on several factors, the two most important being its biological relevance and the nature of the radiation emitted.

Table 1 shows properties of a few of the radionuclides most commonly used in autoradiography.

As already mentioned tritium and carbon-14 are most extensively used owing to the ease with which they can be incorporated into almost any organic molecule. Sulphur-35 is most frequently used in the form of methionine, cysteine, or sulphate while iodine-125 is conveniently used to 'tag' ring-structure amino acids of peptides and proteins.

Specific radioactivities are inversely related to the rate of decay (i.e. half-life). Tritium, carbon-14, and sulphur-35 are beta-emitters. The beta-particles which they produce are electrons of nuclear origin and are given off as a continuous spectrum of energies the maximum and average values being shown in Table 1. Iodine-125 and iron-55 are examples of 'electron-capture' isotopes which stabilize by emitting electrons from the atomic shells (Auger electrons) plus gamma and X-rays. The Auger electrons are produced as discrete line spectra the most important of which are shown in Table 1. It is these electrons which are important for production of the silver grains at the light and electron microscopic levels of autoradiography.

Table 1. *Radionuclides commonly used for autoradiography*

Isotope	Half-life	Useful decay	Principal energy
Tritium	12.26 years	beta	E_{max} 18 KeV; E_{av} 5.5 KeV
Carbon-14	5760 years	beta	E_{max} 156 KeV; E_{av} 50 KeV
Sulphur-35	87.2 days	beta	E_{max} 167 KeV; E_{av} 49 KeV
Iron-55	2.7 years	Auger e	5.5 KeV
Iodine-125	60 days	Auger e	2.9 KeV; 0.8 KeV

The gamma-photons emitted are of considerably higher energy and will contribute only to autoradiographic images produced in films used for macro work (see below) and are most useful in radiometric studies.

The half-life, particle energies, specimen thickness, and emulsion type and thickness determine exposure time. These parameters will be considered in greater detail in later chapters.

2.3. Emulsions

Since specific emulsions and their use are described in relation to particular applications and levels of autoradiography in the following chapters, only a few general remarks are needed here.

2.3.1. Emulsion thickness

It should be remembered that there are two basic ways to use emulsions in autoradiography. If the emulsion thickness is at least equal to the path length of emitted particles (alpha, beta) a 'track' autoradiograph is produced where many silver grains reveal the path of the particle.

In practice, track autoradiography is used only where the particles maintain sufficient energy to travel in straight lines as with alpha-emitters. The relatively small atomic mass of electrons causes them to lose energy rapidly during their passage through emulsions. This renders them easily deflected such that their paths, as revealed by silver grains in their tracks,

Table 2. *Films which can be used for macroautoradiography*

Film	Isotope	Characteristics
Kodak X-Omat (Eastman Kodak)	^{14}C and isotopes of higher energy	Suitable for direct or scintillation exposure High sensitivity
Industrex C (Kodak U.K.)	^{14}C and higher energies	Medium sensitivity Good contrast
Scopix CR3 (Agfa-Gevaert)	^{14}C and higher energies	Medium sensitivity Good contrast Coated on one side only
Ultrofilm (LKB)	Tritium	Specifically for ^{3}H Single-coated, no anti-scratch layer 12 × more sensitive than X-Omat; 64 × sensitivity of Industrex C
Hyperfilm-^{3}H (Amersham)	Tritium	Similar to Ultrofilm
Hyperfilm-betamax (Amersham)	^{14}C and higher energies	Fine grain High sensitivity Coated on one side only

are often tortuous making the position of their origin much less clear than in the case of alpha-particles.

Therefore, in the interest of good resolution (i.e. minimizing image spread), the majority of practical autoradiography is of the 'grain density' type where each particle gives rise to one (or fewer) silver grains. Most of the silver grains are kept relatively close to the decay source by minimizing the thickness of the specimen and emulsion layer according to the type of autoradiograph required.

2.3.2. Physical form of the emulsion

With the exception of macroautoradiography, emulsions may generally be applied to the specimen in liquid form although a preformed layer is sometimes useful in microscopic autoradiography.

The choice and use of nuclear emulsions is determined by several factors. Thus to minimize exposure time, emulsion sensitivity (equivalent to film speed in photography) should be high. This is achieved through increased crystal size with or without extra chemical sensitization. It must be remembered that this 'speed' is likely to lead to increased background fog and, if crystal size is greater, there will also be some loss of spatial resolution.

Photographic emulsions possess halide crystals with a range of sizes to give a broad greyscale in the final picture. Conversely, nuclear emulsions need to provide a detector with a uniform response and so the crystal diameters are relatively similar.

Where a plausible choice of emulsion exists, it makes sense to choose the

Table 3. *Nuclear emulsions classified according to the Ilford scale of sensitivity*

Manufacturer	Mean crystal diameter (μm)	Sensitivity					
		0	1	2	3	4	5
Ilford	0.27						G5
Ilford	0.20	K0	K1	K2			K5
Ilford	0.14					L4	
Kodak (UK)	0.20			AR-10			
Eastman–Kodak	0.34						NTB-3
Eastman–Kodak	0.29			NTB			
Eastman–Kodak	0.26				NTB-2		
Eastman–Kodak	0.05			129-01			

Ilford emulsions are available as gels, or coated on to glass plates or polyester films or as pellicles.

Kodak AR-10 available only as stripping film on glass plates.

Eastman–Kodak emulsions are available in gel form only.

Ilford L4 and Kodak 129-01 are the emulsions most suitable for EM autoradiography.

product which is manufactured locally. As well as reducing cost, this tends to avoid the problem of background exposure during transit and storage.

Table 2 shows some films commonly used in macroautoradiography and Table 3 shows some nuclear emulsions suitable for light and electron microscopic autoradiography.

3 Macroautoradiography

3.1. Autoradiography in separation techniques

3.1.1. Chromatography

Although staining methods are the most straightforward way to identify spots on chromatograms, the ease with which they can be detected is often limited by the amount of material available. Autoradiography offers greater sensitivity when specific activities are sufficiently high. Paper chromatograms can be apposed to a sheet of X-ray film and clamped between two boards. Light but even pressure is achieved by tightening a bolt and wing-nut at each corner. As ever the length of exposure depends upon activity, energy of emission and film sensitivity. Sensitive films such as Kodak X-Omat will give an acceptable image for 20 nCi per cm^2 of ^{14}C after 10 h exposure. For this type of film an appropriate safelight would be Kodak GBX-2 or Ilford NX-914 (with a 15 W bulb).

Processing should be as follows:

1. Develop with GBX (Kodak) or Phenisol (Ilford), at 20°C for 4 min
2. Stop-bath (acetic acid 1 per cent)
3. Fix in Hypam (Ilford) 2 min
4. Wash in running water 20 min minimum
5. Dry at room temperature or in a warm air cabinet

Thin layer plates can be subjected to autoradiography by a similar contact method. It is often advantageous to distribute the pressure evenly on both film and plate by placing a layer of foam rubber at the back of both prior to clamping. It is important that the X-ray film should on no account be permitted to rub or slide against the plate as this can both sensitize the film and damage the plate.

3.1.2. Gel electrophoresis

Cellulose acetate or polyacrylamide gel electrophoretograms may also be processed for macroautoradiography. Gels must be carefully dried (in air or vacuum) before applying film in order to minimize cracking or distortion.

Quantification of bands (or spots) can be by densitometry (see Section

10

3.3) although a more reliable approach is to elute radiolabel from given quantities of gel, plate, or paper and count appropriately.

Recently, a beta-scanning microprocessor-controlled system has been introduced for rapid quantification of polyacrylamide and agarose gels, TLC plates, or nitrocellulose filters (Automated Microbiology Systems Ltd., Grundy Building, Somerset Road, Teddington, Middlesex TW11 8TD, UK). The system may be used with most isotopes other than tritium and can produce a digitized image in minutes rather than days. This image can then be quantified by the available software.

3.1.3. Problems with tritium in chromatograms and electrophoretograms

The low beta energies of tritium combined with the specimen density and the anti-scratch layer present on the surface of most X-ray films ensure that autoradiographic sensitivity is very low. Exposure times are therefore often unacceptably long especially for biochemists who tend to work to a much shorter time-scale than cytochemists.

Dipping chromatograms in, or spraying or impregnating with, liquid emulsions offers only marginal improvements and increases the risk of elution of activity from spots.

Immersion of chromatograms in liquid scintillators causes beta-particles to interact with the scintillator to produce light photons. This process has been variously termed 'scintillation autoradiography', 'fluorography', and 'beta-radioluminescence'. In fact the improvements in efficiency have not been remarkable except where lower exposure temperatures (down to $-79°C$) have been used. In addition to the latter measures it is possible to achieve a further order of magnitude of sensitivity by controlled brief pre-exposure of X-ray film to light to increase the rate of formation of stable latent images.

3.2. Plant macroautoradiography

It is possible to show the passage of assimilated or sythesized nutrients in plants by macroautoradiography. For example, roots of tree seedlings can be immersed in a solution containing ^{32}P phosphate for several hours. Alternatively the aerial portion of such a seedling can be isolated in an atmosphere containing ^{14}C CO_2 for some hours.

For such irregular specimens the best way to ensure good apposition is to use an air-tight plastic bag which can be evacuated. Upon exposure and development, radio-phosphate can be seen to have translocated to the leaves, with greatest density in younger leaves. Conversely, ^{14}C labelled products of photosynthesis have been shown to reside in all parts of the plant including roots. Such experiments have been used to establish the acquisition by mycorrhizal fungi of photosynthetic material from the host plant.

3.3. Whole body autoradiography

Whole body autoradiography is used widely in the chemical and pharmaceutical industries to assess toxicity, distribution, and fetal transfer of drugs and other chemicals. Animals have been prepared in a variety of ways for application of tissues to the film such as freezing followed by milling or sawing to reveal the tissues of interest. Despite this, the most common approach is the use of cryosections of whole frozen animals.

The development and refinement of the method is largely due to Ullberg and his co-workers from the Department of Toxicology at the University of Uppsala, Sweden.

Of course, it is possible to assess the anatomical distribution of label by radiometric counting of tissue samples at autopsy. However, since whole body autoradiography allows examination of labelled tissues *in situ*, the smallest organs can be identified and monitored for their radioactive content.

3.3.1. Animals used for whole body autoradiography

Young adult mice are most commonly used for convenience and cost. Rats are also frequently used while more 'exotic' species such as monkeys, newborn pigs, and pike have occasionally been studied by this method. Comparison of distribution of radiochemicals in different animal species can often reveal differences in metabolism and serve as a salutary warning when attempting to extrapolate the results of animal experiments to humans.

Good animal husbandry is important, especially for long-term studies. For example, animals should be kept separate in metabolic cages to minimize the risk of re-ingestion of labelled excreta by licking. This problem can be further reduced by the use of collars.

3.3.2. Administration of compounds

Although radiochemicals may be given orally, intraperitoneally, topically, subcutaneously, or intramuscularly, the majority of studies use the intravenous route and examine distribution from a few minutes to a few hours after injection (but see Section 2.1.1).

Rats and mice are conveniently injected via the tail vein using some form of restrainer if the animal is not anaesthetized. When acute distribution (e.g. between 2 and 10 min) is required, the animal should be anaesthetized before injection. A chloroform jar can be used for convenient anaesthesia. If a barbiturate anaesthetic is used in a long-term study, its possible effects on hepatic metabolism should be considered.

3.3.3. Freezing and embedding the animal

A good general cryogen is hexane cooled with dry ice to $-75°C$. When the interval required after dosing has been reached, the animal, deeply anaesthetized or killed by anaesthetic overdose, is immersed into the freezing fluid. To do this, rats and mice can be held by the tail, although some workers prefer to use metal clamping devices which help to avoid the tendency of the abdominal contents to collapse on the diaphragm when the animal is inverted. Heat transfer through biological tissues is a slow process and an adult mouse takes at least 5 min to freeze in a hexane/CO_2 mixture, whereas a rat may take up to 25 min.

Most workers like to embed the carcass for sectioning in carboxymethyl cellulose (CMC—wallpaper paste). This can be done directly on to the microtome tage, using a mould with detachable sides.

3.3.4. Sectioning, mounting, and freeze-drying

Essentially, a whole body cryostat comprises a large, deep freezing chamber with a motorized sledge-type microtome. Instruments may be obtained commercially from Slee (London), the Bright Instrument Co. (Huntingdon), and PMV (Stockholm). Figure 2 shows the interior of a whole body cryostat from PMV.

Before sectioning begins the carcass must equilibrate to the temperature of the cryostat ($-20°C$). The carcass is then trimmed until a plane is reached which contains the organs of interest. Adhesive tape (e.g. 3M 800 or 810), slightly wider and longer than the section to be taken, is placed on to the frozen specimen and care taken to eliminate air bubbles. A section (typically 20 μm thick) is cut and retrieved by slowly advancing the 'block' to the knife (which may be stainless steel or tungsten tipped). The leading edge of the supporting tape is lifted above the knife edge and the adhering section guided upwards and away.

For freeze-drying, the tape with its section is supported across a wire frame which can be stored in a rack held in the cryostat at $-20°C$. Freeze-drying is normally complete within 48 h.

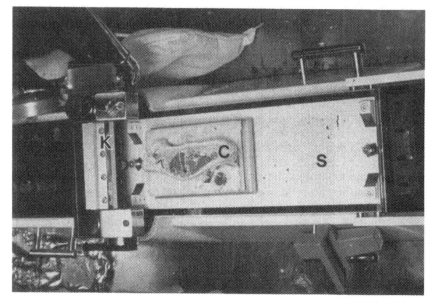

Fig. 2. View of the interior of a whole-body cryostat with the lid opened. Large hydraulic stage (S), trimmed carcass set in carboxymethyl cellulose (C), and knife (K) are visible

3.3.5. Exposure

The freeze-dried sections, which are very hygroscopic, are transferred in a box with a dry atmosphere to the darkroom. The tape (with its section) is placed against an appropriate film (see Table 2) under safelighting as recommended by the film manufacturer. Good contact between section and film is essential and this can be achieved using a photographic roller. For quantification, a radioactive scale may be placed on the film (Section 3.3.8). Once the sections are mounted and suitably labelled the films are replaced in the supplier's envelopes. For emitters such as ^{3}H and ^{14}C the envelopes may be stacked prior to placing in a press for exposure in a light tight container. For higher energy isotopes, such as ^{32}P and ^{125}I, specimens should be stored singly and well separated. Exposure temperatures in the range -10 to $-20°$C are generally used. Length of exposure is a function of the parameters already mentioned and is normally in the range 1 to 4 weeks.

3.3.6. Development of whole body autoradiographs

After exposure, the sections must be removed from the film before they can be developed. The tape should be peeled slowly from the film to avoid static which will cause unwanted blackening. Some workers find that separation under water helps to prevent this problem. Occasionally, certain batches of tape will prove difficult to remove from the film. This problem can be avoided by dusting the section and tape with talcum powder before applying the film.

Developing schedules are recommended according to film type. The protocol listed in Section 3.1.1 is of general use for whole body auto-radiography. There are many possible developers—Kodak D19 being widely available and, in our experience, suitable for most films. Figure 3 shows a typical whole body autoradiograph.

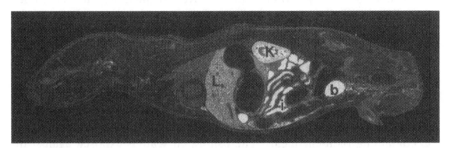

Fig. 3. Whole-body autoradiograph of a rat 40 min after intravenous injection of a tritiated analogue of somatostatin. Note the radiolabel is being eliminated (secreted) by both the liver (L) and kidney (K) so that the contents of intestine (i) and bladder (b) are also labelled. The autoradiograph has been printed in reverse contrast so that the most intensely labelled tissues appear whitest

3.3.7. Use of the section

More often than not, radiolabel is sufficiently widely distributed through the animal tissues to leave on the film a fairly comprehensive image of the anatomy present in the section (Fig. 3). However, identification of certain 'hot spots' may require reference to the section. The natural colour difference of organs, especially in thicker sections (20 μm plus), may be enough to identify organs of interest but, if not, sections can be stained as follows:

1. Fix the sections (and tape) in 10 per cent buffered formol (pH 7.2) for 2 min
2. Rinse in tap water
3. Stain in Erlich's haematoxylin 2 min
4. Rinse in tap water
5. Differentiate in 1 per cent hydrochloric acid
6. Blue in 1 per cent NH_4OH or tap water
7. Counterstain in 1 per cent eosin for 30 sec
8. Rinse in tap water

9. Dehydrate by 1 min immersions in 70 per cent, 90 per cent, and 2 × absolute ethanol
10. Mount tape and section between large format slide and cover-glass using Euparal (Raymond Lamb)

Sections treated in this way can be viewed either macroscopically or microscopically under low power (if 20 μm or less in thickness).

3.3.8. Quantification of whole body autoradiographs

The most accurate way to obtain quantitative information from whole body autoradiographs is to cut a section adjacent to the one to be used for autoradiography which is thicker (perhaps up to 500 μm). From the adjacent section, portions of tissues can be punched out, weighed, and oxygen-combusted or dissolved in M NaOH before scintillation or gamma-counting.

Less time-consuming, and more detailed, is densitometry. Cross *et al.* (1974) have described such a quantitative method using a 'step-wedge' scale made from fixed X-ray film impregnated with serial dilutions of ^{3}H or ^{14}C containing radiochemicals. The calibrated scale is exposed against the same sheet of film as the experimental section. After development, a light-sensitive probe is used to compare transmitted light values for section and scale.

Although useful, this method has been criticized for several reasons, the most important of which is its inability to take into account differential tissue absorption of ^{14}C beta-particles, in particular by mineralized tissues.

Further reading

Cross, S. A. M., Groves, A. D., and Hesselbo, T. (1974). A quantitative method for measuring radioactivity in tissue sections for whole-body autoradiography. *Int. J. Appl. Rad. Isotopes.* **25**, 381–6.

Curtis, C. G., Cross, S. A. M., McCulloch, R. J., and Powell, G. M. (1981). *Whole-body autoradiography.* Academic Press, London.

Ullberg, S. (1977). The technique of whole body autoradiography, *Science Tools* (Special Issue) LKB-Produkter AB, Stockholm.

4 Light microscopic auto-radiography—fixable substances

This chapter deals with the preparation of specimens in which the radiolabel is presumed to be retained in its *in vivo* sites and in which steps have been taken to remove all unincorporated label (see further reading list for suitable fixation and embedding methods for specific classes of labelled compounds).

Generally, the emulsion is applied to a section of embedded tissue. Also, cultured cells or cells in suspension which, for example, have been incubated with the labelled DNA precursor thymidine and dried on to slides can be prepared for autoradiography using this technique.

4.1. Choice and preparation of slides

It is impossible to generalize about microscope slides and their treatment prior to autoradiography. Slides with a ground-glass end on one side can be convenient when working in safelighting conditions, for labelling purposes and as a reminder of which side has the section. Normally, however, it is preferable to use scratched labelling on plain slides to avoid any unwanted chemical additions to the specimen.

Depending on the source of the slides, various degrees of cleaning will be required. Chromic acid cleaning may be needed for some types while 90 per cent ethanol will suffice for others.

Since sections (and emulsions) adhere well only to certain types of microscope slide, many workers prefer to 'sub' slides by dipping them into a high-grade gelatin solution (0.5 per cent w/v). The experience of the author has been that subbing is essential only when using stripping film (see below).

4.2. Section type

4.2.1. Paraffin wax sections

Nowadays, paraffin sections around 4–6 μm thick are used for autoradiography where the emphasis is on minimizing exposure time (for isotopes other than tritium) whilst retaining adequate, if not outstanding, histological quality. For quantitative work, the variation in section thickness

17

should be monitored. This can be done by thoroughly mixing a non-polar radiolabel with paraffin wax and cutting serial sections from blocks of this. The sections are then dissolved (e.g. with toluene) and counted in a liquid scintillation counter.

After dewaxing, the sections should be kept wet to avoid distortion when applying the emulsion.

4.2.2. Resin sections

When fine histological detail is required and some sacrifice of radiolabel (and therefore exposure speed) is acceptable, resin sections give a more satisfactory result than paraffin wax. Many histology laboratories are now equipped for routine production of semi-thin (0.5–1.0 μm) resin sections of large tissue blocks. Where this is the case, 'Ralph' type knives are generally used for sectioning. Typically, the resins used are methacrylates which are sufficiently hydrophilic to permit some penetration of most aqueous stains.

Alternatively, the electron microscopist may wish to carry out a 'rapid' sampling survey of his small tissue blocks. These are usually embedded in epoxy resins which have greater stability than acrylics in the electron microscope beam. It may well be advantageous to collect the semi-thin sections in series with ultra-thins whilst working at the ultratome. Larger blocks can be taken for semi-thin sectioning before trimming to size for electron microscopy. Small platinum wire loops are used to transfer sections from the knife-bath to the slide which can be marked on the reverse side to indicate the position of the sections. It is usually best to have the sections about one-third from the end of the slide for reasons of emulsion economy and convenience during processing.

4.3. Preparative methods for light microscopic autoradiography

4.3.1. Stripping film

To all intents and purposes, the only commercially available example of stripping film is the fine grain product AR 10 manufactured by Kodak UK (see Table 3). This product incorporates a 5 μm emulsion layer on a 10 μm gelatin layer which in turn is coated on to a glass plate. Stripping film is suitable for just about any LM autoradiographic experiment with the commonly used isotopes. The product is essential for quantitative work with isotopes such as ^{14}C and [35]S. This is because these isotopes emit beta-particles whose maximum range in organic matter (tissue section plus emulsion) is between 10 and 80 μm. Hence, the majority of electrons emitted by ^{14}C and ^{35}S pass straight through the emulsion. The latter must therefore be of uniform thickness for each atom of isotope to have an equal probability of producing a latent image.

Since, in the case of tritium, the maximum range of the beta-particles is between 1 and 3 μm, they are all absorbed within the specimen or emulsion layer. Hence, all have a chance to produce latent images irrespective of emulsion thickness.

Owing to increased use of liquid emulsions, stripping film has diminished in importance as a general-purpose emulsion. Nevertheless, in spite of reduced demand and the concomitant problem of maintaining manu-facturing skills, Kodak, to their credit, have continued to make this product available. This has been in response to those autoradiographers who have pointed out the unique nature of AR 10. It must be assumed that the future availability of stripping film is constantly under review and cannot be taken for granted.

4.3.2. Application of stripping film

For Kodak AR 10, safelighting should be with a Kodak Wratten 1 filter (25 W bulb, 1.2 m minimum distance).

For floating out the stripping film, a fluid bath is used which can be filled either with distilled water or, when many weeks exposure are anticipated, with a 20% (w/v) sucrose solution containing 0.01 per cent (w/v) potassium bromide. The temperature of the bath should be between 20 and 23°C.

AR 10 is supplied in boxes containing 3 packs of 4 glass plates in each of which the plates are stacked as 2 pairs with the emulsion layers facing inwards but separated by a plastic spacer (Fig. 4a). Storage should be at 4°C.

The film should be allowed to come to darkroom temperature (20–22°C) for a few hours before being unpacked and used. A sharp scalpel is used to score a margin along all 4 sides of the plate (about 5 mm). The film is then scored to divide the area into rectangles of about 2.5 × 5.0 cm (Fig. 4b). The scalpel is used to lift the corner of each rectangle which can then be taken between finger and thumb (which must be dry!) and stripped away from the plate. The ease of separation of film from plate depends on the relative humidity of the darkroom which for this procedure should be about 65%. If the relative humidity is too high the film will be reluctant to leave the glass and tend to tear. If the relative humidity is too low the film will detach almost without help but will be inclined to curl making floating out difficult.

To float out on to the bath the film must be inverted so that it is emulsion side down. Care must always be taken not to wet the fingers since this will make handling of subsequent emulsion rectangles difficult.

Within 3 minutes, imbibition of water by the emulsion and gelatin layer causes the area of the film to increase about 1.4-fold. From this it follows that the emulsion and gelatin thickness during exposure are about 3.6 and 7.1 μm respectively. Once the film is no longer expanding it should be picked up on a specimen-bearing slide before it becomes waterlogged and sinks. The slide is positioned below the floating film such that the labelled sections are directly covered. The slide is lifted up one edge before the other so that the ends of the emulsion strip wrap around the back of the slide (Fig. 4c).

The slides are placed in a rack. Drying should be in air and can take from one to many hours depending on darkroom humidity.

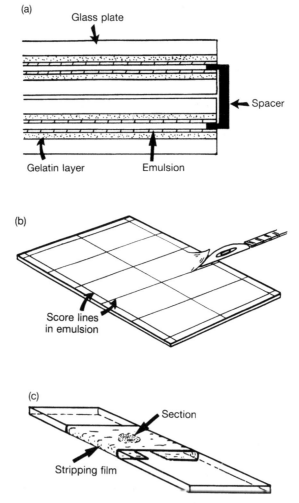

Fig. 4. (a) Diagram to show how four sheets of Kodak AR-10 stripping film are packed together. (b) Method for stripping AR-10 from its glass plate. (c) Diagram of a slide bearing a labelled section covered with stripping film

4.3.3. 'Liquid' emulsions (emulsions in gel form)

If the investigator wishes to do qualitative work or quantitative work with tritium only, the emulsions supplied in gel form have several advantages over stripping film. They are available in a range of crystal sizes and sensitivities (Table 3) and are more economic. They are variously manufactured in North America and Europe and are thus more 'locally' available. In addition, they have some considerable advantage over stripping film in relation to staining (see Section 4.6).

For LM autoradiography of the commonly used isotopes (e.g. ^{3}H, ^{14}C, ^{35}S) the Eastman–Kodak NTB ('nuclear track') and Ilford K range are most suitable. Better resolution and lower background are offered by those emulsions with small crystal size and low sensitivity. Even so, most users are concerned primarily with obtaining an adequate result in the shortest practicable exposure time. With this in mind, it is probable that the most generally useful gel emulsions for LM autoradiography are NTB-2 and K5 and their use is described here.

The physical properties of the two emulsion ranges are very different. Ilford now market a derivative of K5 called K5D in North America only, intended for those workers used to handling the NTB range. Photographically, it is identical to K5.

4.3.4. Preparation of LM autoradiographs with Kodak NTB-2

The recommended safelighting filter for the whole NTB range is a Wratten 2. The emulsion is supplied in a plastic bottle in solid gel form so it must be liquified before use. This is done in a water bath at 43–45°C for 30–45 min. Slides can be dipped into the stock bottle, but if used in this way the stock is at risk of becoming contaminated. Consequently it is advised to transfer gently some of the emulsion to a flattened dipping jar or, if unavailable, a Coplin jar—preheated in the water bath.

Any bubbles which form should be removed by repeated gentle dipping and withdrawal of clean slides. When a bubble-free layer is achieved on the dipped blank slides the specimens may be coated with emulsion. The experimental slides are immersed to a depth which ensures the sections are covered with emulsion and then slowly and steadily withdrawn. Excess emulsion is drained off the vertically held slide with a tissue which is then used to wipe clean the reverse side of the slide. The slides are placed horizontally on an ice-cold metal surface (emulsion side up) to allow the emulsion to gel, which under conditions of reasonable humidity, will be complete within 30 minutes.

Precise emulsion thickness will depend upon speed of withdrawal of slides and length of time in the vertical position. Generally, this method produces emulsion layers which are no more than 4 μm thick.

Before being put away to expose, the slides and their emulsions should be perfectly dry. This can be done by leaving the slides in racks in the darkroom overnight. Figure 5 shows a home-made light-tight box used by the author through which air can be circulated and which avoids the need to keep the darkroom locked up overnight.

Fig. 5. Home made exposure box for LM autoradiographs showing exterior (a) and interior (b) views. There are light-tight vents (v) in the base and lid so that drying can be accelerated by placing a small fan below the box

4.3.5. Preparation of LM autoradiographs with Ilford K5

The safelighting recommended by Ilford for their range of gel emulsions is F904 (orange) but Kodak Wratten 2 filters are also safe.

All Ilford emulsions are supplied in brown glass bottles and as 'worm'—like shreds. If there is a solid gel in the bottle, this indicates that the emulsion has melted and it should be returned as suspect.

For dipping slides, K5 needs to be diluted 1:1 as follows: preheat in a 43–45°C water-bath a dipping jar containing 12 ml distilled water, a measuring cylinder marked visibly at 12 ml, and a 100 ml glass beaker. Use plastic forceps or spoon to take emulsion from the stock bottle so as to fill the beaker approximately one-quarter. Allow this to melt for 10–15 min and then gently pour into the measuring cylinder up to the 12 ml mark. This 12 ml of emulsion is then added to the water in the dipping jar. The diluted emulsion is stirred very slowly for 2 min and allowed to stand for a further 5 minutes such that bubble formation is minimal. Some workers recommend the addition of 1% glycerol to this mixture as a plasticizer although it is probable that the plasticizer included by Ilford is sufficient to minimize background fog produced by drying stress.

Testing, dipping, and subsequent treatment of experimental slides is as for the NTB-2 method and the final K5 layer will be in the region of 3 μm thick.

There are home-made and commercially-available dipping machines designed to standardize the withdrawal of slides from the emulsion and thereby standardize the emulsion layer. For light microscopic work, however, all that is needed to achieve a satisfactory consistency is a little practice and patience.

4.3.6. Chemography

Chemography is a sensitization response in the emulsion crystals not produced by radiation. Positive chemography is the production of silver grains by chemical interaction of the specimen with the emulsion. It is controlled by identically prepared non-radioactive specimens which strictly should have been labelled with the 'cold' equivalent of the experimental radiolabel.

Negative chemography is the chemical desensitization of emulsion crystals by the specimen such that formation of latent images by radiation is prevented or reversed. Its presence is revealed by totally fogging (in white light) one or two slides from each batch at the beginning of exposure. If, upon development any part of the tissue section is visible through the blackened emulsion, negative chemography has occurred.

If either positive or negative chemography persists it may be eliminated by interposing an inert layer between section and emulsion such as a thin evaporated carbon layer. This, however, will make subsequent staining difficult.

4.4. Exposure of LM autoradiographs

Needless to say, it is important not to 'put all your eggs in one basket'. Hence duplicate batches at least of all experimental slides should be put up for exposure.

Slides should be placed in black plastic boxes containing self-indicating silica gel packed in a way that prevents its dust from contacting the slides. The slide boxes are sealed with black insulating tape and left in a refrigerator (at 4°C) to expose. Many types of slide box which are not black are also not light-tight and should be avoided unless sealed inside black plastic bags.

4.5. Development

The following are suitable processing schedules for the three emulsion types described in this chapter (safelighting as already described).

Kodak AR 10
1. Develop in undiluted Kodak D19 4 min at 19°C
2. Stop in running tap water 30 sec, 18–21°C
3. Fix in 30% (w/v) sodium thiosulphate 10 min, 21°C
4. Wash in gently running tap water 5 min, 18–21°C
5. Wash in two further changes of distilled water 5 min each, 18°C
6. Air-dry slowly in a dust-free atmosphere

Kodak NTB-2
1. Develop in Kodak D-163 (UK) (1:1) or Dektol (USA) (1:1) 2 min, 16°C or develop in Kodak D19 (1:1) 4 min, 16°C
2. Stop in distilled water 10 sec, 16°C
3. Fix in 30 per cent sodium thiosulphate 5 min, 16°C
4. Wash in distilled water 5 min, 16°C
5. Air-dry slowly in a dust-free atmosphere

Ilford K5
1. Develop in Ilford ID19 (= Kodak D19) (1:1) 4 min, 19°C or develop in Ilford Phenisol (1 + 4) 5 min, 19°C
2. Stop in 0.2 per cent acetic acid bath 30 sec, 19°C
3. Fix in 30 per cent sodium thiosulphate 10 min, 21°C
4. Wash in running tap water 15 min, 18–20°C
5. Slowly air-dry in a dust-free atmosphere

4.6. Staining

In practice, there is little to recommend the prestaining of sections for LM autoradiography. Prestaining greatly increases the risk of chemography and some stains under certain circumstances actually emit photons. Moreover, certain stains are eluted by developing and fixing procedures.

For paraffin sections haematoxylin and eosin is a useful stain combination although staining times should be minimized to avoid excessive uptake by the emulsion gelatin. This is particularly problematic with the gelatin backing layer of AR 10 and is a major reason why the so-called 'liquid emulsions' are often preferred. An important point when staining stripping film autoradiographs is to keep all solutions at around 17°C. This helps to prevent detachment of film from the slide. After staining stripping film preparations, and while the film is still wet, excess film on the back of the slide should be removed. Air drying is preferable to alcohol drying in autoradiography.

Autoradiographs of semi-thin plastic sections are conveniently stained with the metachromatic dye toluidine blue (1 per cent in 1 per cent boric acid). Gentle warming at 37°C of a drop of stain over the section permits uptake of the stain by the tissue even in hydrophobic epoxy resin sections. Preparations may be differentiated in 50% ethanol before washing and air drying.

4.7. Preparation for microscopy

Mounting of autoradiographs prepared with stripping film is considered unnecessary by many workers since the 7 μm gelatin layer acts as a 'surrogate' coverglass. Unfortunately, this layer is easily scratched even when wiping off immersion oil with lens tissue.

Mounting media of the DPX type have been implicated in long-term disappearance of developed grains. This might be a problem if stored autoradiographs need to be examined again.

4.8. Examination, quantification, and micrography of LM autoradiographs

Although transmitted bright field microscopy reveals much of the auto-radiographic image to the observer's eye, there are often many grains, particularly of the background, which the observer does not readily register. Figure 6 shows how dark field illumination can reveal these 'extra' silver grains. This principle becomes very important in the automated analysis of LM autoradiographs.

The simplest and cheapest approach to quantification of LM autoradiographs is visual counting of grains in the bright field mode using an eyepiece graticule. This method is undoubtedly tedious and time-consuming when compared with automated photometry systems generally used with dark field or reflected epi-illumination (Dörmer 1972; Goldstein and Williams 1971).

In visual grain counting, one is measuring the number of silver grains present, which in turn are deemed to have a numerical relationship to the number of decays taking place in a tissue compartment. On the other hand, it has been shown that reflectance measurements assess only total silver coverage and cannot distinguish grain numbers from development effects on grain size (Rogers 1972). While this does not negate the value of automated methods it underlines the need to standardize preparative conditions in specimens which are to be compared numerically.

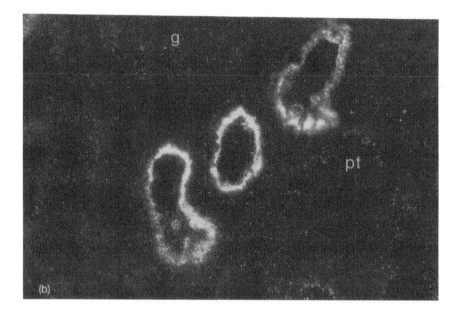

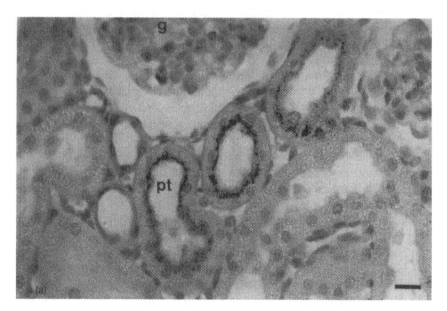

Fig. 6. Stripping film LM autoradiograph of perfusion fixed rat kidney cortex 3 min after intravenous injection of a tritiated adrenocorticotrophin analogue. Silver grains are easily visible in the apical zone of three proximal tubule profiles (pt) when viewed with bright field illumination (a). When dark field illumination is used (b) other proximal tubules (pt) and glomerulus (g) are seen to contain radiolabel as well. Scale bar = 20 μm

A particular problem of photomicrography of LM autoradiographs is that objective lenses above 25× do not have sufficient depth of focus to allow coincident sharp imaging of silver grains and tissue section. This problem can be largely overcome by double exposure of focused images of first the tissue and then the grain layer.

Further reading

Dörmer, P. (1972). In *Microautoradiography and electron probe analysis*, (ed. U. Luttge), Springer-Verlag, Berlin.

Goldstein, D. J. and Williams, M. A. (1971). Quantitative autoradiography: an evaluation of visual grain counting, reflectance microscopy, gross absorbance measurements and flying spot microdensitometry. *J. Microscopy*, **94**, 215–39.

Rogers, A. W. (1972). Photometric measurements of grain density in autoradiographs. *J. Microscopy*, **96**, 141–53.

Rogers, A. W. (1979). *Techniques of autoradiography*. (3rd edn). Elsevier, Amsterdam.

Singh, R. N. (1986). How to increase the effective depth of focus of the light microscope for photography. *Stain Tech.* **61**, 17–20.

5 Light microscopic auto-radiography—diffusible substances

In autoradiographic terms, a working definition of a 'diffusible' (or 'soluble') substance is one which is water or lipid soluble, can diffuse freely through cells and tissues, and cannot be totally retained in its *in vivo* site via chemical fixation (see Section 2.1.3).

5.1. Examples of diffusible substances

The following is a non-exhaustive list of chemicals which are usually found to behave as diffusible substances:

Steroids (oestrogens, glucocorticoids, mineralocorticoids)
Glycosides (digoxin, ouabain—steroid related)
Sugars (when not involved in biosynthesis)
Fatty acids (when not involved in biosynthesis)
Opiates (morphine, diprenorphine—excluding endorphins)
Ions (Na^+, K^+, Ca^{++}, Mg^{++}, Cl^-, PO_4^{--})
Many pharmaceuticals which are not peptides or amines (the latter are often fixable with aldehydes).

Some ions may be precipitated by ions of opposite charge and thus be retained in what is hoped to be their *in vivo* sites. A well-known example of this is pyroantimonate which although showing a preference for calcium will also co-precipitate with other cations such as sodium and magnesium. The latter, however, is a rather specialized approach rarely used in auto-radiography, more often in conjunction with X-ray microanalysis in the electron microscope.

Very few methods have emerged to cope with diffusible substances. Furthermore, whatever the radiochemical concerned, each method has been based on freezing the labelled specimen, i.e. 'cryofixation' as opposed to chemical fixation. Hence it can be seen that whole-body autoradiography is a method eminently suitable for diffusible substances. At the other end of the scale, diffusible substance autoradiography at the EM level is not yet fully evaluated nor easily performed (see further reading list). Thus it is on the LM level that we shall concentrate in this chapter.

5.2. Physical chemistry of the freezing of biological specimens

It is important to have at least a basic appreciation of this in order to understand what might be happening to the distribution of radiolabelled solutes during the cooling and warming of cells and tissues.

In effect, freezing is the removal of pure water from solution and its isolation into biologically inert foreign bodies—ice crystals. The freezing point of cytosol is usually a little below 0°C but many live cells remain unfrozen (i.e. 'supercooled') down to −15°C even when ice is present in the external medium. This is because cell membranes tend to prevent growth of external ice into the supercooled interior.

If cells are cooled slowly they equilibrate by transfer of intracellular water to external ice. Thus a process of 'heterogeneous nucleation' takes place in which relatively few ice crystals grow. On the other hand, rapid cooling brings about formation of numerous small intracellular ice crystals, a process of 'homogeneous nucleation'.

Whichever type of freezing occurs, it can be seen that formation of ice crystals will have the effect of displacing and concentrating solutes (including radiochemicals). This artefact, however, will be reduced when nucleation is homogeneous as a result of rapid freezing.

It is important to note that the lower the temperature, the slower will be the rate of ice crystal growth until, at −130°C, no further ice crystal growth can take place.

5.3. Physical chemistry of the warming of frozen biological specimens

When the temperature of frozen cells or tissues is raised, the most important phenomenon for the autoradiographer is 'recrystallization'. This is the preferential growth in the solid state of large ice crystals at the expense of small ones.

At very low temperatures (about −100°C), recrystallization is relatively slow. At temperatures nearer the melting point (−60°C and above) ice crystals grow large rapidly.

Therefore, it is vital to remember that all procedures up to the end of exposure are carried out in the absence of an aqueous phase. Moreover, the final autoradiograph is of a freeze-dried section which has been rehydrated by fixation, development, fixing, and staining which re-anneals any ice crystal voids created by the freezing or warming processes. The investigator must therefore ask himself whether the final grain distribution represents *in vivo* label or a mosaic of label trapped between ice crystal voids within the section during exposure!

5.4. Methods for autoradiography of diffusible substances at the LM level

As with so many techniques, those for autoradiography of diffusible substances have tended to be used in the locality of the inventor. Thus two major methods were developed simultaneously in the early 1960's. That of T. C. Appleton has been used mostly in Europe while the Stumpf and Roth technique has found its greatest following in North America. Here, a working account of the Appleton method is given followed by a brief summary of two alternatives.

5.4.1. The method of Appleton

Tissues should be frozen as rapidly as possible by dropping them into a cryogen, preferably at its melting point such as Arcton 12 cooled in liquid nitrogen or in nitrogen slush (liquid N_2 reduced to a melting solid at $-210°C$ by placing in a vacuum of 10^{-2} Torr). If it is necessary to store specimens, this is best done in liquid N_2.

Once fully frozen, tissues are mounted on to cryostat microtome chucks using a medium such as Tissue-Tek (carboxymethyl cellulose and glycerol) or water and then rapidly freezing either by contact of the chuck with dry ice or immersion in a liquid cryogen. It will be obvious that sectioning must never get to the level of the tissue which was in contact with the mounting medium and in which diffusion may have occurred. Before sectioning, tissues must be allowed to come to the temperature of the cryostat at around $-30°C$.

Since the main methods for diffusible substance autoradiography preclude contact of the specimen with liquids until the end of exposure, emulsions must be preformed on a slide or coverglass and applied dry to the specimen. Stripping film may be used as described in Section 4.3.2 with an important modification, that in this case it must be floated out emulsion-side uppermost since, when mounted, the cryosections will lie above rather than beneath the emulsion. Alternatively, emulsion layers can be made with 'liquid emulsions' as in Sections 4.3.4 and 4.3.5.

Some workers prefer to coat coverglasses rather than slides with emulsion so that in the final autoradiograph the section is viewed conventionally, lying below the emulsion. In practice this is not important and purely a matter of personal preference. Once dry, the coated slides (or coverglasses mounted onto slides) are stored in light-tight boxes with silica gel inside the chamber of the cryostat.

Cryotomy is carried out in the darkroom under white lighting (not fluorescent because of the risk of 'afterglow' after switching off). Cryosections between 5 and 8 μm are suitable and their morphological features can be checked by thawing a test section on to a plain slide, fixing in 5 per cent/95 per cent (v/v) acetic ethanol and staining in toluidine blue (1 per cent in 1 per cent boric acid). Once a zone of interest in the specimen is reached, go to safelighting and accommodate visually before proceeding. Sections which adhere either singly or in ribbons to the knife edge, may be picked up by moderate but firm pressure of a pre-chilled emulsion-coated slide or coverglass. If adhesion to the emulsion proves to be a problem, it may be useful to pre-chill the emulsion in a separate chamber or container maintained at a higher sub-zero temperature (e.g. -15 to $-20°C$).

If, after applying the emulsion, there is any sign of hoar-frost around the section this is an indication that the section has thawed throughout its complete thickness and this may have permitted diffusion of the radiolabel. All such preparations must be discarded but do become very infrequent once a competent level and speed of working have been achieved. Figure 7 shows cryostat sections being mounted on to an emulsion-coated slide.

Fig. 7. Mounting cryostat sections on a chilled emulsion-coated slide under safelighting for the Appleton technique

Exposure of slides or coverglasses in light-tight boxes with silica gel is in the cryostat or other freezing chamber maintained at a similar temperature. During the exposure period the sections completely freeze-dry.

After exposure, preparations may be processed according to the following schedule:

1. Fix (histologically) by immersion in either acetic ethanol (5 per cent/ 95 per cent v/v) at 17°C, 3 min, or in 10 per cent buffered formol.

2. Rinse in two changes of distilled or gently running tap water at 17°C, 4 min.

Development and fixing can be carried out as detailed in Section 4.5, depending upon emulsion type. It is important in this case however, not to allow the temperature of any processing fluid to exceed 17°C or to use agitation at any time. This helps to avoid detachment of sections from the wet emulsion.

After air-drying, sections can be stained as described in Section 4.6. The method of mounting will depend upon the original support for the emulsion (slide or coverglass). It should be noted that the hazards mentioned in Section 4.7 apply equally to autoradiographs of diffusible substances.

There can be no pretending that the Appleton method is as technically easy or free of artefacts as conventional 'fixed-label' autoradiography. Even so, with care and practice it can be made to produce reliable results which limit diffusion of any radiolabelled substance within the ability of the light microscope to discriminate the distribution of the autoradiographic image. Figure 8 shows an autoradiograph obtained by the Appleton method.

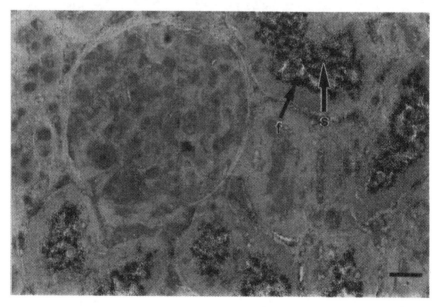

Fig. 8. LM autoradiograph produced by the Appleton method using stripping film. The experiment is identical to that in Fig. 6 except that instead of using perfusion fixation the tissue was excised and frozen in Arcton 12 (at −156°C). The silver grains indicate label resorbed by the proximal tubules much of which is aldehyde-fixable (f) plus soluble label present within the filtrate (s). Scale bar = 20 μm

5.4.2. The method of Stumpf and Roth

In this method, cryostat sections are also used but are cut at a somewhat lower temperature ranging between −30 and −85°C according to the section thickness required. The sections are collected from the cryotome knife using a mounted needle or the bristles of a fine brush and transferred to a small 'polyvial'. The latter in turn is transferred to a freeze-dryer. After thorough freeze-drying the sections are allowed to come to room temperature and placed in a desiccator.

Microscope slides are dipped with a liquid emulsion (originally NTB-3), air-dried, and stored with desiccant.

Section mounting must be carried out with great dexterity and at a low relative humidity (not above 25 per cent). The dried section, which is hygroscopic and fragile, is placed on a Teflon support. An emulsion-coated slide is placed over the section and the Teflon support and slide are pressed briefly together with forefinger and thumb after which the Teflon falls away with the section adhering to the emulsion.

An advantage of this method is that freeze-drying reduces the chance of chemography. On the other hand, the sections are fragile, difficult to handle, and the method is technically more difficult than Appleton's with more apparatus needed. Most importantly, the laboratory conditions required are difficult to achieve in some places such as the UK where humidity tends to be excessive and air conditioning primitive or absent.

5.4.3. The method of Stirling and Kinter

These collaborators introduced an interesting alternative to cryosections. In their original publication, rings of hamster intestine incubated with ^{3}H-galactose were frozen, freeze-dried, fixed in osmium tetroxide vapour, and embedded in Araldite containing silicone fluid.

It is disquieting that 25 per cent of radiolabel from 1 μm sections was lost to the knife-boat water in spite of the hydrophobicity afforded by the silicone. Nevertheless, subsequent studies from these authors showed apparently specific localization of ^{3}H-galactose in pre-incubated entero-cytes of normal patients but not in those with galactose malabsorption. A notable feature of the Stirling and Kinter method is the superior histological quality of the preparations.

There is no doubt that in all methods used for diffusible substance autoradiography, histological quality is compromised especially since tissues which have been freeze-dried exhibit impaired staining. Despite this, practice, perseverence, or even beginner's luck can bring acceptable and useful results.

Further reading

Appleton, T. C. (1964). Autoradiography of soluble labelled compounds. *J. R. Microsc. Soc.* **83**, 277–81.

Appleton, T. C. (1986). Resolving power, sensitivity and latent image fading of soluble-compound autoradiographs. *J. Histochem. Cytochem.* **14**, 414–20.

Baker, J. R. J. and Appleton, T. C. (1976). A technique for electron microscopic autoradoigraphy (and X-ray microanalysis) of diffusible substances using freeze-dried fresh frozen sections. *J. Microscopy,* **108**, 307–15.

Stirling, C. E. and Kinter, W. B. (1967). High-resolution radioautography of galactose-^{3}H accumulation in rings of hamster intestine. *J. Cell Biol.* **35**, 585–604.

Stirling, C. E., Schneider, A. J., Wong, M-D., and Kinter, W. B. (1972). Quantitative radioautography of sugar transport in intestinal biopsies from normal humans and a patient with glucose–galactose malabsorption. *J. Clin. Invest.* **51**, 438–51.

Stumpf, W. E. and Roth, L. J. (1965). Dry-mounting high-resolution autoradiography. In *Isotopes in experimental pharmacology* (ed. L. J. Roth), pp. 133–43. University of Chicago Press, Chicago, Illinois.

6 Autoradiography of receptors

Receptor mapping by the autoradiography of radioligands is a specialized use of the autoradiographic principle and currently its fastest growing application.

Since the majority of ligand–receptor interactions are reversible, the technique depends basically on the use of cryosections as described in Chapter 5 for the minimization of diffusion. 'Ligand' is a general term for endogenous hormones or pharmacological agonists and antagonists which are capable of interaction with receptors.

Early attempts at diffusable substance autoradiography to identify receptor distribution used *in vivo* labelling of experimental animals. However, the technology which has produced the current explosion in receptor autoradiography is based on *in vitro* labelling of tissue cryosections and is primarily attributable to Dr Michael Kuhar of the Johns Hopkins Medical School, Baltimore, USA.

The *in vitro* labelling method has been exploited mostly by neuroscientists interested in the central nervous system since the many ligands which do not cross the blood–brain barrier may then be studied. Other advantages of the technique are:

1. Washing may be used to reduce unbound labelling to low levels
2. Adjacent (serial) sections can be used to compare receptor (or ligand) distributions
3. The effects of nucleotides, ions, or other mediators on binding can be selectively demonstrated
4. The method is much more economic than the *in vivo* alternative with respect to animals and radiopharmaceuticals
5. Early postmortem tissues, including those from humans, can be used to study receptor distribution.

6.1. Methodology

6.1.1. Preparation of animal tissues

Tissues may be excised from the sacrificed animal and rapidly frozen in a cryogen of choice (e.g. isopentane at $-40°C$ or Arcton 12 in liquid nitrogen at $-156°C$). However, experience has shown that light perfusion fixation of the organ of interest with freshly prepared 0.1% formaldehyde in phosphate buffered saline (pH 7.4) gives superior histological appearance to the final section without detectable effect on ligand/receptor binding. Cryostat sections (about 10 μm) are thaw-mounted on acid-cleaned subbed microscope slides stored at room temperature. The sections are allowed to dry on the slide at room temperature for 1 to 4 h to ensure good adhesion. The slides can be stored at $-20°$C for up to 20 days prior to use.

6.1.2. Incubation of tissue sections

Sections are brought to room temperature for at least 45 min before incubation which is normally carried out at 22°C. The appropriate composition of the incubation medium is determined by the tissue and radiochemical. Tris buffer (about 180 mM, pH 7.6) is commonly used and the radioligand (usually tritiated or radio-iodinated) is used in the picomolar or nanomolar range. Clearly, temperature, time, and dose-dependency for optimal labelling should be assessed for each combination of tissue and radioligand.

When the tissue of interest has a high lysosomal content, such as the kidney, there are some advantages in the inclusion of a protease inhibitor such as phenyl-methylsulphonal fluoride (10 μM) in the incubation medium. Incubation can be carried out by immersion of slides in a jar (see Fig. 9) although some workers prefer to pipette the incubation medium on to slides which are held horizontally in a moist chamber.

The specificity of labelling can be monitored by controls in which sections are pre-incubated or co-incubated in an excess (in the micromolar range) of an unlabelled competitive antagonist. Unbound label is removed by two rinses (up to 15 min each) in incubation buffer alone.

6.1.3. Radiometry of sections

Sections to be used for radiometric counting are best wiped or scraped off the slides (before they are dry) into a counting vial. Sections containing iodine-125 are gamma-counted whilst those with tritiated ligands are dissolved and mixed with a suitable cocktail (e.g. Packard ES 299 or LKB Lumagel) and scintillation-counted. Figure 10 shows an hypothetical (but typical) time–course binding curve for a labelled ligand in which specific receptor binding is calculated by subtracting non-specific counts (those obtained in the presence of excess 'cold' antagonist) from total bound counts. The ordinate values can be expressed as counts per minute (c.p.m.) per section or as moles per wet weight of tissue.

The specific binding characteristics of a given ligand and its respective receptor have traditionally been described by Scatchard analysis of tissue homogenates. Similar analysis can be performed with cryosections by

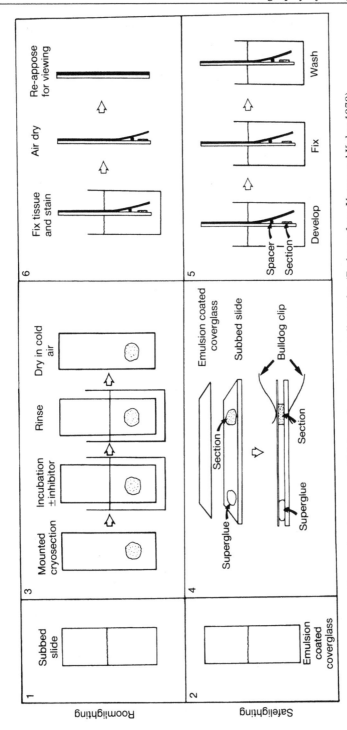

Fig. 9. Flow diagram showing the essential steps of receptor autoradiography. (Redrawn from Young and Kuhar 1979)

counting bound ligand following incubation with a range of ligand concentrations. Figure 11 shows a typical Scatchard plot and the parameters of binding which can be deduced from it. The greater the dissociation constant (K_d) the smaller becomes the slope of the plot which denotes a lower binding affinity (i.e. smaller bound/free ratio).

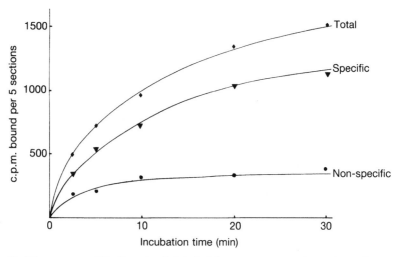

Fig. 10. Time course of binding of radiolabelled ligand to receptors in sections. Non-specific counts represent activity bound in the presence of excess unlabelled ligand and are subtracted from total counts to obtain the specific figure

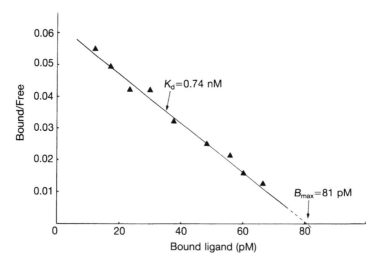

Fig. 11. Scatchard plot derived by incubating sections in different ligand concentrations and determining the concentration bound per mg protein. The single line illustrated indicates a single population of receptors. If the slope of the plot changes, this indicates a second dissociation constant (K_d) and thus a mixed population of receptors

6.1.4. Preparation of autoradiographs of radioligand binding

As depicted in Fig. 9, the original description involved acid-washed cover-glasses dipped in Kodak NTB-3 (1:1). Others have used Ilford K5 (1:1.5). The air-dried and desiccated coverglasses are apposed to the section-bearing slides (under safelighting) using Superglue (Loctite Corp.). Up to six such preparations are held fast by a bulldog clip at the section end. Exposure duration at 4°C with desiccant is determined by trial develop-ment of replicate batches of slides. For development and subsequent processing the coverglass and slide are held apart using a spacer (Fig. 9). Developer should be chosen suitably for the emulsion used (see Chapter 4).

If organs were initially perfused with formaldehyde, it is possible to use a histochemical reaction or histological stain to aid interpretation of the final autoradiograph. Finally, the slide and coverglass are permanently reunited by removal of the spacer and addition to mounting medium (e.g. DPX or Permount). Examination of the autoradiographic image is usually carried out by dark field microscopy whilst the stained tissue sections are viewed conven-tionally in the bright field mode. More recently, the method has been adapted for 'semi-macro' applications by substituting the emulsion-coated coverglass for Ultrofilm (see Table 2). Figures 12 and 13 show autoradiographs of receptor binding using a K5 coated coverglass and Ultrofilm respectively.

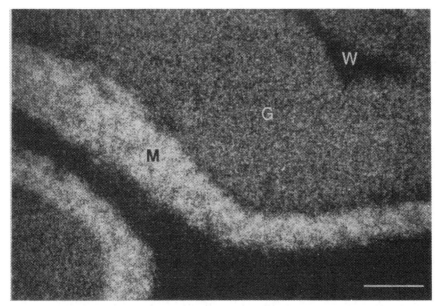

Fig. 12. LM darkfield autoradiograph (using Ilford K5) of B receptors to tritiated gamma aminobutyric acid (GABA) in rat cerebellum. The grain densities reveal the greatest con-centration of receptors in the molecular layer (M), with fewer receptors in the granular layer (G), and white matter (W) respectively. Scale bar = 300 μm. (Kindly supplied by Professor N. G. Bowery)

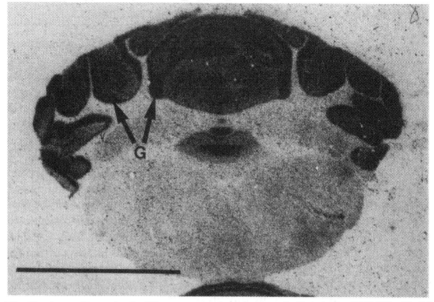

Fig. 13. Autoradiograph of a 10 μm cryosection of rat cerebellum (using LKB Ultrofilm) to show GABA-A sites which are most concentrated in the granular cell layer (G). Scale bar = 5 mm. (Kindly supplied by Professor N. G. Bowery)

6.2. Functionality of ligand-bound receptors

The major criterion by which a binding site can be regarded as a receptor is when interaction of that receptor with a specific ligand can be shown to elicit a cellular response.

It is now known that many hormone-receptor interactions bring about phospholipase C-dependent cleavage of phosphatidyl inositol bis-phosphate into the second messenger substances inositol 1,4,5 tris-phosphate (IP_3) and diacyl glycerol. A few investigators are now beginning to show that increased grain densities in autoradiographs of radioligand receptor binding can indeed be correlated with IP_3 production in parallel experiments.

6.3. Quantification of receptor autoradiographs

Since there is increasing evidence that the grain densities in receptor auto-radiographs satisfy expected binding kinetics and functional criteria, many workers now wish to quantify their results. The major problem is to convert grain densities to a molar value for ligand binding per unit area of tissue.

Hence some form of standard is required against which experimental grain densities can be calibrated (compare with the 'step-wedge' used for whole body autoradiography, Chapter 3).

Polymer-based microscales containing ^{3}H, ^{14}C, or ^{125}I are commercially available (Amersham International).

A less expensive alternative is to manufacture one's own standards by homogeneously radiolabelling tissue paste with serial dilutions of radiolabel which can be frozen and sectioned as an intact tissue. As with the use of all scaled standards, the differential tissue self-absorption of the soft beta particles of tritium or Auger electrons of iodine-125 must be taken into account.

From a size point of view, it is apparent that the commercial microscales are most useful in the calibration of LM microdensitometry. Home-made tissue paste scales are larger and more conveniently used for quantification of macro images with computer-based image analysers (e.g. Quantimet 920 or 970; Magiscan 2a). These modern (and expensive) systems can be used to subtract non-specific grain densities from total grain densities to leave a digitized image representing specific receptor labelling. Such images can in turn be colour-coded using appropriate standards (paste or otherwise). Figure 14 shows a black and white representation of a colour-coded auto-radiographic image.

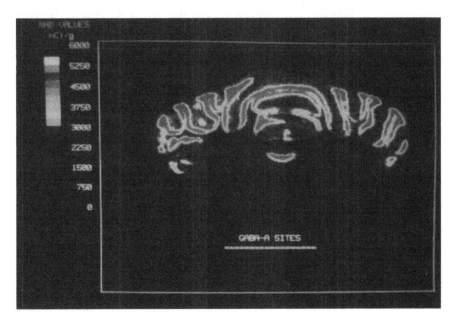

Fig. 14. Black and white half-tone representation of a colour-coded equivalent of the specimen in Fig. 13 used for quantification and produced on a Quantimet 920 image analyser. (Kindly supplied by Professor N. G. Bowery)

6.4. Hybrid DNA autoradiography and receptor labelling

It is now a fact that autoradiography to detect cDNA for *in situ* hybridization has been superseded by more convenient and higher resolution probes such as immuno-gold or streptavidin-biotin gold. Nevertheless, many investigators still favour autoradiography for this purpose and receptor coding mRNA in developing (gestational) tissues can be revealed by *in situ* hybridization with radiolabelled (^{35}S, ^{32}P, ^{125}I) cDNA probes.

Further reading

Clark, C. R. and Hall, M. D. (1986). Hormone receptor autoradiography: recent developments. *TIBS*, **11**, 195–9.

Gusterson, B. A., Neville, A. M., Baker, J. R. J., and Christian, R. A. (1981). Autoradiographic localisation of ^{3}H-diprenorphine in the Syrian hamster. *Diagnostic Histophathology*, **4**, 209–13.

Hunt, S. P., Manthy, P. W., and Ninkovic, M. (1986). The localisation of neuropeptide receptors (Third Int. Symposium on Autoradiography, Sheffield, U.K.). *Proc. Roy. Microsc. Soc.* **21**, 274.

Kuhar, M. J. (1983). In *Handbook of Chemical Neuroanatomy Vol. 1: Methods in Chemical Neuroanatomy* (eds A. Bjorklund and T. Hokfelt), pp. 398–415. Elsevier, Amsterdam.

Wilkin, G. P., Hudson, A. L., Hill, D. R., and Bowery, N. G. (1981). Autoradiographic localisation of GABA-B receptors in rat cerebellum. *Nature*, **294**, 584–7.

Young, W. S. 3rd and Kuhar, M. J. (1979). A new method for receptor autoradiography: (^{3}H) opoid receptors in rat brain. *Brain Res.* **179**, 255–70.

7 Electron microscopic auto-radiography—preparative methods

Autoradiography is carried out at the EM level to relate distribution of radiolabel within tissues to the ultrastructural detail obtainable from examination of ultra-thin sections in the transmission electron microscope.

Before proceeding to prepare EM autoradiographs it is especially important to ensure that the fixation schedule used has retained sufficient radiolabel within the specimen to produce an adequate grain density in a reasonable exposure time (see Chapter 2). This is because the extreme thinness of both section and emulsion layer renders the EM level of autoradiography slow to produce latent images (and thus silver grains) as will be demonstrated in the next chapter.

It is beyond the scope of this book to describe details of fixation and EM preparative schedules suitable for particular experimental conditions.

One special circumstance which should be mentioned is when cells in suspension have been labelled. It is strongly recommended that, once fixed, such cells are diluted and 'embedded' in a gel such as serum albumin cross-linked with glutaraldehyde as described by Bullock and Christian (1976). This has the dual advantage that repeated centrifugation and resuspension steps are avoided during subsequent preparation. Also, if the cells are sufficiently well dispersed within the gel, the radiation field of each cell is separated from that of its neighbours. This can have important implications for analysis of the final autoradiographs.

There are so many layers in the final autoradiographic 'sandwich' that very high standards of preparation and cleanliness must be maintained at each step.

7.1. Ultramicrotomy

Because of the sampling hazards inherent in electron microscopy, every effort must be made to ensure that sufficient blocks are used to be fully representative of the labelled tissue. To this end also, the area of block surface sectioned should be as large as possible. Hence a diamond knife is to be preferred to glass since it will allow larger and more sections to be cut. This is important for the necessary levels of replication for multiple exposure batches for each experimental treatment.

Under ideal conditions (i.e. for best resolution) the thinnest possible

43

sections should be produced—these give a silver/grey interference colour (about 50 nm). In practice, it is usually better to forego the advantage of resolution and opt for sections in the gold range of thickness (about 100 nm). This is because thicker sections contain more radiolabel so that exposure time is reduced. In addition, the extra thickness will give better contrast which is always at a premium since it is reduced in the autoradiograph by the presence of the semi-opaque gelatin layer of the emulsion.

Subsequent handling of the sections depends on which of the two major methods for emulsion application is used.

7.2. The loop technique

The application of nuclear emulsion to ultra-thin sections using a wire loop was first described by Caro and Van Tubergen (1962), and this has been widely used in variously adapted forms. The method described here is based on that recommended by Williams (1977) and modified according to the experience of the author.

7.2.1. Treatment of sections

Formvar or collodion coated (grey interference colour) gilded copper grids (200 mesh size) are used. The copper offers enough rigidity to withstand the several handling steps involved while the gold-plating offers a surface to which the nuclear emulsion is less sensitive.

Sections are picked up on to the rough surface of the grid from below the water bath surface. This is more difficult than retrieval by contact from above but tends to avoid creases in the sections.

Once dry, sections are stained by inverting the grids on drops of stain as follows:

1. Uranyl acetate (5 per cent aqueous solution)—10 min
2. Reynolds lead citrate—5 min

There is little doubt that contrast is more easily achieved by pre-staining than by post-staining of autoradiographs. Anxiety that use of a uranium salt introduces unwanted radioactivity into the specimen is largely unwarranted since the alpha tracks which are only very occasionally produced by ^{235}U in the final autoradiograph cannot be confused with the single silver grains which are of experimental interest (Fig. 15).

There are two plausible disadvantages to the prestaining of EM autoradiographs. The first is the possibility of positive or negative chemography. This is controlled as described in Section 4.3.6 and avoided by evaporating a thin layer of carbon (about 5 nm) on to the grids. (It is worthwhile at the same time to carbon coat some support film-carrying grids

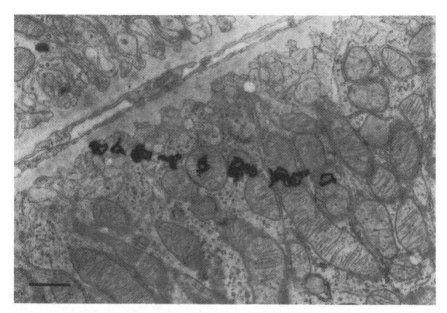

Fig.15. Alpha track produced in the plane of a monolayer of Ilford L4 as seen in the EM (Microdol X development). The alpha-particle was produced by ^{235}U in the uranyl acetate used to stain the section. Scale bar = 1 μm

without sections—these can then be used for testing the uniformity of the emulsion layer before its application to experimental grids.) The second disadvantage to pre-staining is connected with the fact that the commonly used developer, Kodak D19, is relatively efficient in terms of grain production due to its high pH (about pH 10). This means that, under certain circumstances, Reynolds lead citrate, which is also soluble at high pH, may be partially leached by such a developer and re-deposited within the section in a non-anatomical way. This problem can be avoided by *en bloc* staining with lead aspartate which is used at pH 5.5 and is thus protected from dissolution at alkaline pH.

7.2.2. Preparation of grids for application of emulsion

Once the sections have been stained and carbon-coated, the grids are placed shiny-side down on a cork (about 1 cm apical diameter) bearing double-sided Sellotape in which a hole has been punched and whose diameter is just less than that of the grid (Fig. 16).

7.2.3. Choice of emulsion

Clearly, the most appropriate emulsions to use for EM autoradiography are

Fig. 16. Corks with double-sided Sellotape in which a hole is punched whose diameter is just less than that of a grid bearing emulsion-coated specimens. The corks are conveniently stored in PVC tubing in which a hole has been drilled to permit drying

those with the smallest crystal diameters so that resolution in terms of grain size and distance from the decay source can be optimized. In practice, the preferred product for use with the loop technique worldwide has been Ilford L4. This is mainly because of limitations of the alternatives. For example, Sakura HR-2 is commercially available only in Japan. Also, Eastman–Kodak's 129-01 (Table 3) and its predecessor NTE, were developed specifically with the flat substrate technique in mind and do not readily form a stable 'membrane' on a wire loop as does Ilford L4.

7.2.4. Preparation and application of L4 with the loop technique

The following is a check-list of items which are needed in the darkroom:

MATERIALS

- A wire loop (diameter 3 cm) made from 21 gauge nickel/chrome or platinum wire set in a wooden or glass handle
- A sliding weight balance
- A water bath (of the histology 'floating out' non-agitating type) set at 45°C
- Clamp for a 250 ml beaker

- An ice bath
- Glass stirring rod
- Jar of Ilford L4 emulsion and plastic forceps
- Measuring cylinder containing 250 ml glass distilled water
- A 2% stock solution of the detergent Mannoxol OT (dioctyl sodium sulphosuccinate—available from Polysciences)
- A 1 ml graduated pipette
- The specimens, mounted on corks (and the 'dummy' test grids)
- Black slide boxes, sachets or gelatin capsules containing self-indicating silica gel, short lengths of PVC tubing about 5 cm long which can accept the corks and which have a hole drilled in the side, reel of black insulating tape, scissors
- Suitable safelighting—Ilford F902 (yellow), Ilford F904 (orange), Kodak Wratten 2 (red)

METHOD

1. Using the graduated pipette, withdraw 0.4 ml of the Mannoxol OT and add to the cylinder containing 20 ml distilled water. Place the 250 ml beaker on the pan of the sliding weight balance and adjust until the balance is swinging freely. Now add 10 g to the beam and switch to safe-lighting only. Using the plastic forceps add pieces of emulsion to the beaker until the balance is swinging freely. Now add the water/Mannoxol to the beaker and clamp so that the lower third of the beaker is immersed in the water bath.

2. When the emulsion has begun to melt and 'dissolve' it may be gently stirred (about 1 revolution per second for 10 seconds every minute). After 20 min the beaker should be removed from the water bath and placed in the ice bath. For exactly 3 min the emulsion is stirred constantly at the previous rate taking care not to create bubbles.

3. At the end of this time the beaker is placed on the bench and the emulsion allowed to 'semi-gel' slowly at room temperature. To ensure even gelling the emulsion must again be stirred for at least 10 seconds in every minute.

4. In the absence of the detergent, the emulsion gels rapidly in about 5 min and soon becomes too viscous to use. The Mannoxol has the effect of prolonging the gelling time to at least 15 min. More importantly, it allows the ideal semi-gelled condition to be maintained for up to 1 h and thus permits the coating of many specimens.

5. The correct condition for use is reached when the emulsion flows slowly in the beaker (like double cream) and has a convex meniscus.

6. To test the emulsion, the beaker is held at 45 degrees (around the rim so that hand heat does not re-melt the emulsion) and the wire loop is immersed horizontally. This is easiest if the loop is bent at 45 degrees to the handle. The loop is then turned through 90 degrees and withdrawn

vertically and slowly to avoid making bubbles in the viscous emulsion which, once present, are difficult to remove.

7. When held up to the safelight, the ideal 'loop' should be of uniform opacity (Fig. 17a) and be stable on the wire. If the membrane is uneven or 'flows' on the wire (Fig. 17b) the emulsion is not yet sufficiently gelled.

8. Before applying the emulsion membrane to the mounted experimental grids, two or three test grids should be coated and examined in the electron microscope. It is important that small condenser 1 settings and a well defocused condenser 2 be used to minimize volatilization of halide crystals in the beam. Figure 18a shows an acceptable emulsion 'mono-layer' and Fig. 18b an uneven and unacceptable crystal distribution.

9. The Mannoxol has one further advantage. If even monolayers are consistently elusive it is possible (with Mannoxol present) to dry the emulsion on the loop. The length of time needed for drying depends on the darkroom temperature and humidity and can take between 2 and 6 min. As the membrane dries, a series of interference colours appears (easier to see in yellow than orange or red safelighting). When a patch of gold is present in the membrane, apply this to the grid. If the membrane appears shiny, a short breath over the grid is necessary to ensure good adhesion of the extremely dry emulsion layer. The emulsion so formed on the grid is usually an overlapped monolayer but is of uniform thick-ness which is vital for quantitative work.

10. The optimum method for preparation of Ilford L4 is affected by many variables such as darkroom temperature (ideally not above 20°C), relative humidity (which should not exceed 50 per cent) and emulsion batch. The method described here is currently in use by the author and gives satisfactory results. If the reader is consistently unsuccessful in the

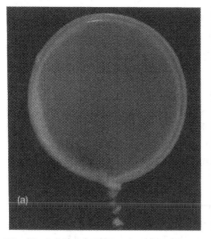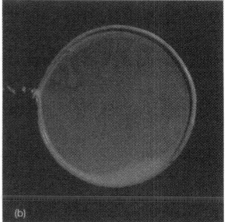

Fig. 17. (a) Stable L4 loop, acceptable for coating specimens and (b) emulsion which is still flowing on the loop and cannot be used

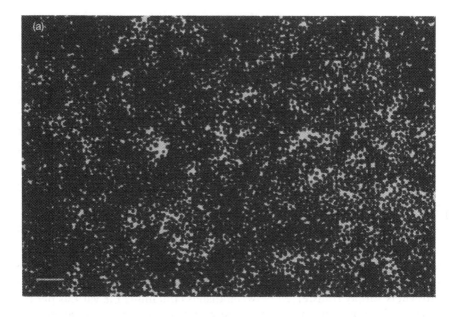

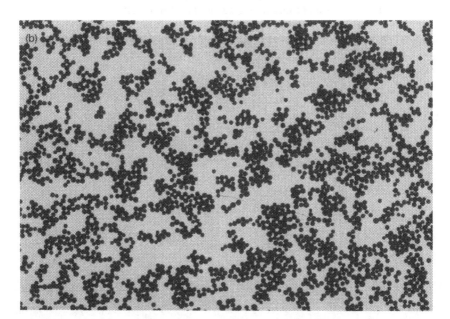

Fig. 18. (a) Electron micrograph of a packed 'monolayer' of Ilford L4 produced by the loop method and suitable for application to experimental sections. (b) Broken layer of L4 produced by applying the loop bearing over-gelled emulsion to the grid. Scale bar = 1 μm

use of the method such that uniform monolayers cannot be obtained, he/she is advised to check darkroom conditions or to vary slightly the emulsion dilution.

11. Once all the experimental grids have been coated with emulsion and the corks placed in to the tubing (one in each end) the preparations are put away to expose. For exposure, black plastic boxes containing sachets or capsules of silica gel are used and sealed with black tape. The specimens should be prepared at least in duplicate so that correction for over- or under-exposure can be made if necessary (see below). The best temperature for exposure is 4°C.

7.2.5. Development of EM autoradiographs made by the loop technique

Two basic approaches to the handling of specimens during development, etc., are possible. The first and most convenient is to mount the specimens in a platform into which holes have been drilled as shown in Fig. 19. This platform is simply transferred between dishes containing the various photographic solutions. Unfortunately, since many surfaces are in contact with the solutions, there is a chance that grids will become soiled.

Fig. 19. Perspex developing platform for EM autoradiographs made by the loop method

Slightly more laborious but generally much cleaner, is to transfer grids between drops of solutions on sheets of dental wax or Parafilm as in routine EM staining. The relatively bright safelighting permissible for L4 makes this method quite feasible.

The two commercially available developers from Kodak, D19, and Microdol X are widely used and very suitable for Ilford L4 under most circumstances. Other developers, such as phenidon (which produces concentric grains), paraphenylene diamine (which produces composite grains), and elon-ascorbic acid (with or without gold latensification) are described by Williams (1977). The Kodak developers both give the familiar large convoluted silver grains when viewed in the electron microscope. Some workers believe that by using a developer which gives smaller grains they can get better resolution. This is largely a fallacy and it should be borne in mind that most of the published data on EM autoradiographic resolution (i.e. grain spread around the source of decay) were derived using D19 or Microdol X.

The major difference between D19 and Microdol X is due to their different pH values, being about 10 and 7.8 respectively. This means that at a given temperature and duration of development, D19 develops approximately 3 times as many latent images as Microdol X.

The following is a suitable development schedule for L4 coated grids (all solutions should be freshly made, filtered, and used at 19 or 20°C):

1. D19 or Microdol X 3 min
2. Stop in distilled water (Microdol X) or 1 per cent acetic acid (D19)
3. Fix in 30 per cent sodium thiosulphate 4 min
4. Wash in distilled water 2 × 5 min
5. Finally, wash grids in a jet of distilled water from a Pasteur pipette (× 2)—dry the grids on filter paper

Figure 20 shows an EM autoradiograph, prepared by the loop method, of a lymphocyte whose cell membrane has been labelled with iodine-125.

7.3. The flat substrate technique

The main alternative to the loop method of emulsion application is the so-called 'flat-substrate' technique which was introduced by Salpeter and Bachmann (1964). Originally, molten emulsion was dropped on to collodion coated slides bearing ultra-thin sections but this was rapidly superseded by dipping the slides in the emulsion. Hence the method has become known as the 'dipping technique'. One of its advantages is that it is suitable for use with almost any liquid emulsion type. The version of the method described here is derived from that used by Dr Robert Gould of the Institute for Basic Research in Developmental Disabilities, Staten Island,

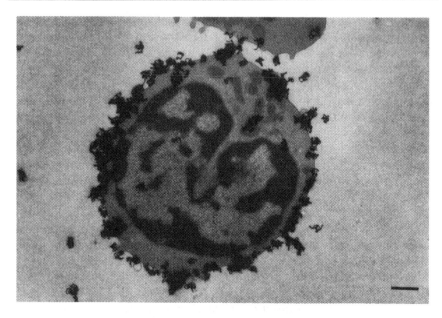

Fig. 20. EM autoradiograph produced by the loop method using Ilford L4. The specimen is a human peripheral lymphocyte whose cell membrane has been radio-iodinated by ^{125}I NaI catalysed by lactoperoxidase. The hydrogen peroxide substrate was provided endogenously by the glucose/glucose oxidase reaction. Scale bar = 1 μm

New York, USA. The method is reliable and does not require any kind of dipping machine.

7.3.1. Preparation of collodion-coated slides

MATERIALS
- Microscope slides
- Lens paper
- Beaker (150 ml)
- Collodion (0.4–0.8 per cent, i.e. 4–8 g pyroxylin/100 ml n-butyl acetate)
- Paper towels and slide drying box
- Tape

METHOD

1. Pour 0.4–0.8 per cent collodion into the 150 ml beaker, using a fume cupboard
2. Clean some slides with lens paper
3. Dip each slide into the beaker of collodion so that 65–75 per cent of the slide is coated
4. Allow excess collodion to drain back into the beaker

5. Allow the slides to dry with the dipped ends on a paper towel and the tops resting on the box (Fig. 21a)

6. After drying (10–20 min), place slides in the box in the fume cupboard

7. Mark the non-dipped end of each slide with a piece of tape (Fig. 21b)

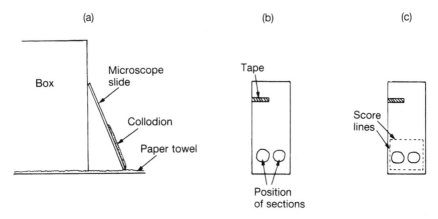

Fig. 21. The flat substrate method of emulsion application for EM autoradiography: (a) drying the collodion support film; (b) marking the location of ultra-thin sections; (c) scoring around the autoradiographic 'sandwich' prior to floating off and mounting on to grids. (Diagrams used by permission of Dr R. Gould)

7.3.2. Putting ultra-thin sections on to slides

1. Place two drops of distilled water on the slide, each approximately 2.5 cm from the dipped end (Fig. 21b)

2. Transfer sections (silver to pale gold) to the slide with a fine platinum loop (32–36 gauge). Apply between 2 to 10 per drop depending on the size of the sections

3. Remove excess water with filter paper. Mark with a circle the position of sections on the underside of the slides using a felt marker or wax pencil (must be non water-soluble)

4. Prepare 4–5 slides per block of tissue

7.3.3. Pre-staining the sections

MATERIALS

- Two large glass Petri dishes with lids (cover the lid of one Petri dish with aluminium foil)
- Pieces of glass in each dish to create a platform (used glass knives are suitable)

- Uranyl acetate in a light tight container (1 per cent in 0.1 M maleate buffer, pH 4.2). Do not use alcohol in this stain because it causes the film to stick to the slide
- Reynolds lead citrate
- Pasteur pipettes

METHOD

1. Place wetted filter paper in one large Petri dish
2. Arrange glass triangles on to the filter paper so that four slides can be placed in the Petri dish
3. Pipette drops of uranyl acetate, one over each group of sections (two per slide) and place the foil covered lid over the slides
4. After 5 min remove the lid and wash the stain down each slide into an empty beaker (250 ml) with a gentle stream of distilled water. Allow the slides to dry
5. Stain slides in a similar manner with lead citrate in a dry Petri dish (using NaOH pellets) one at a time for 2 min

---

7.3.4. Coating the slides with Ilford L4

MATERIALS

- Water bath at 45°C
- Dipping jar containing 20 ml distilled water and 4 drops of glycerol
- 'Short' cylinder, i.e. 100 ml cylinder cut off at a height corresponding to 50 ml
- Measuring cylinder (50 ml)
- Ilford L4 emulsion
- Spatula
- Light-tight boxes
- Black tape
- Silica gel capsules
- Slide drying rack
- Several clean test slides (collodion-coated)

METHOD

1. The pre-stained slides are coated with an evaporated carbon layer approximately 5 nm thick. The dipping jar and both cylinders are pre-warmed in the water bath. Then, under safelighting, emulsion is transferred (using plastic forceps) to the 50 ml cylinder up to the 25 ml level
2. After 15 min in the water bath the molten emulsion is transferred to the short cylinder until it reaches the 5 ml mark
3. The 5 ml of molten emulsion are then gently poured into the dipping jar containing glycerol and water. Because of its dilution, the emulsion

will remain liquid at room temperature and it is usually easiest to carry out dipping at room temperature

4. A few test slides should be dipped to see if the dried emulsion layer gives a suitable interference colour (purple with L4). If the layer is too thick or thin, additional water or liquified emulsion can be stirred in. The emulsion can then be retested with additional slides

5. Once the correct interference colour is achieved, the slides containing experimental sections can be dipped. As with the test slides, these should be dipped to a constant level on the slide and pulled from the emulsion at a repetitive and uniform rate

6. Blot the bottom of the slide (but not the back) with a tissue and allow to dry on a slide rack for 1 h

7. Place the slides in a light-tight box with silica gel capsules and expose at 4°C

7.3.5. Coating the slides with Kodak 129-01

For slightly higher resolution work, Kodak 129-01 (see Table 3) may be used for the flat substrate technique. Type 129-01 emulsion is in a solid gel form when received. The shelf-life of the stock is best protected by removing chunks of emulsion with a clean plastic spoon and melting only what is necessary for a particular experiment (see Section 7.3.4). Mono-layers of crystals can be prepared by using the emulsion from the bottle appropriately diluted with distilled water. With 129-01 a gold interference colour indicates a monolayer. It is recommended to test dilutions of 1:2, 1:4, or 1:8 (v/v) molton emulsion to distilled water. Whichever of these gives closest to the desired result is determined by darkroom temperature, humidity, and speed of withdrawal of the slide from the liquid emulsion. Figure 22 shows a monolayer of 129-01.

7.3.6. Development of flat substrate EM autoradiographs (Ilford L4)

Usually, several slides (4–5) are prepared from each piece of embedded material. The first slide of each series should be developed at an appropriate interval.

MATERIALS
- Kodak D19 or Microdol X—undiluted
- Fixer—Kodak Rapid Fix or 25% sodium thiosulphate
- Slide drying rack

METHOD
1. Prepare containers of developer, distilled water, and fixer at 20°C

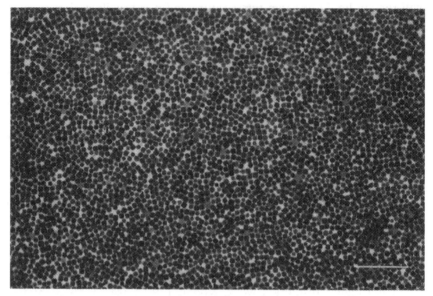

Fig. 22.   Electron micrograph of an even monolayer of Kodak 120-01. Scale bar = 1 μm

2. Under appropriate safelighting, place slides in the staining racks
3. Develop for 2 min at 20°C
4. Rinse for 30 sec in distilled water
5. Fix for 2 min
6. Wash for 2 min in distilled water
7. Wash for a further 2 min in fresh distilled water (during transfers, avoid unnecessary agitation of slides which might cause sections to separate from the collodion).
8. Place the slides on a drying rack and allow them to dry thoroughly

7.3.7. Development of EM autoradiographs (Kodak 129-01)

The procedure described in Section 7.3.6 may be used for Kodak 129-01 except that the appropriate developer is Kodak D163 (UK) or Dektol (USA). In both cases the duration of development should be 2 min and dilution should be 1 part developer:2 parts water.

7.4. Transferring the autoradiographs to grids for viewing

MATERIALS
- Crystallizing dish (75 mm high × 150 mm diameter) containing distilled water

- Suction apparatus (small filter funnel attached by tubing to a pump)
- Cu grids (cleaned in chloroform and dried)
- Filter papers (5.5 cm Whatman no. 1 or no. 4)
- Scoring utensil (diamond scriber or blade)
- Hydrofluoric acid (1 per cent)
- Pasteur pipettes

METHOD

1. Score the slides around the area containing the sections (Fig. 21c)
2. Float the film on to distilled water in the dish. If the film does not float off it can be rescored and 1 per cent HF can be dribbled around the score line
3. Once the film has floated on to the water, the sections are visualized (a magnifier may help) and copper grids, dull side down, are placed over them
4. Mark on the filter papers (with a pencil) appropriate information (e.g. embedment no., days of exposure, date of development)
5. Wet a filter paper and pick up over the end of the suction funnel
6. Pick up the film on the filter paper by quickly touching on to the water surface
7. Dry the filter paper (overnight at room temperture—covered) in an oven at 50°C or with warm air from an air blower (hair dryer)
8. The grids are removed and viewed directly or the collodion can be thinned or removed by placing the grids in n-butyl acetate, ethanol, or acetone for 5 min

Further reading

Bullock, G. R. and Christian, R. A. (1976). A novel approach for enzyme histochemical and autoradiographic studies on single cells. *Histochem. J.* **8**, 291–300.

Caro, L. G. and Van Tubergen, R. P. (1962). High resolution autoradiography 1. Methods. *J. Cell Biol.* **15**, 173–88.

Salpeter, M. M. and Bachmann, L. (1964). Autoradiography with the electron microscope. A procedure for improving resolution, sensitivity and contrast. *J. Cell Biol.* **22**, 469–77.

Walton, J. (1979). Lead aspartate, an *en bloc* contrast stain particularly useful for ultrastructural enzymology. *J. Histochem. Cytochem.* **27**, 1337–42.

Williams, M. A. (1977). *Practical methods in electron microscopy: Autoradiography and immunocytochemistry* (ed. A. M. Glauert), Elsevier, Amsterdam.

8 Efficiency

The remaining chapters deal with the evaluation of autoradiographs. Once the initial problems of preparing satisfactory autoradiographs have been overcome, the greater challenge is to extract the maximum information from the preparations. Essentially the objective is to interpolate back from the observed autoradiographic image so that a valid inference can be made about the location and perhaps the quantity of radiolabel within the specimen. Hence, it is important to appreciate those factors which determine the amount of information which can reach the emulsion, its probability of detection, and its geometric relation to the source of decay.

The proportion of disintegrations occurring in the specimen which are recorded by the emulsion layer is termed the 'efficiency' of the system. Efficiency should not be confused with 'grain yield' which is the average number of grains formed by each particle which enters the emulsion.

Efficiency is directly related to the exposure time required to produce satisfactory grain density autoradiographs. The maximization of efficiency is thus more important at the electron microscopic than the light microscopic level of autoradiography which in turn is more important than at the macroscopic level.

8.1. Efficiency in macroscopic autoradiography

X-ray films are several times more sensitive than nuclear emulsions and in general are not limiting for efficiency especially when isotopes of the energy of ^{14}C or higher are used. Relatively few grains will produce visible blackening of the film since crystal diameters are so large and exposure times are often of very few days duration. For routine work, a high sensitivity film such as Kodak X-Omat (Table 2) should be used. Lower sensitivity film, such as Kodak Industrex C, will require about 5 times as much exposure for equivalent blackening but gives better contrast (e.g. for reports or publication) owing to reduced background fog. The latter two film types are coated on both sides of the backing layer which for isotopes of high beta-particle energy like ^{32}P gives added blackening.

Most X-ray films also contain a surface anti-scratch layer which makes work with the difficult nuclide tritium even more problematic. This is because the short range of ^3H beta-particles allows only the upper 2–3 micrometres of sections to register the signal entering the halide layer.

These electrons are seriously impeded by the anti-scratch layer of X-ray films.

Specifically designed to overcome the problem of macroautoradiography with tritium is the product marketed by LKB as Ultrofilm (Table 2). This contains a highly sensitive emulsion, is single coated, and has no anti-scratch layer. This material has proved surprisingly easy to handle and a very similar product, Hyperfilm-3H is now available from Amersham International, apparently to supply the growing receptor autoradiography market (see Chapter 6).

Perhaps the most worrying aspect of efficiency in macroautoradiography is related to differential ^{14}C beta-particle attenuation in sections (see Section 3.8.8). Longshaw and Fowler (1978) have used 'beta-radiography' to evaluate the extent of attenuation of beta-particles in sections used for whole-body autoradiography. This involved a thin sheet of ^{14}C polymethyl methacrylate used as a radiation source. Freeze dried sections of unlabelled rat ranging from 10 to 60 μm were sandwiched between the plastic source and X-ray film. After exposure and development the light image recorded revealed the range of beta-particle attenuation which might be expected as a result of variation in tissue density. The results suggested that sections which were 10 or 15 μm thick underwent uniform attenuation in soft tissues (44–48 per cent) whilst bone absorbed up to 57 per cent of light. At higher section thickness, beta (i.e. light) absorption was greater and more variable with very high values in bone (up to 89 per cent). It follows that the problem of beta-attenuation in tritium-labelled tissues is much reduced and section thickness is of little importance since all the beta particles reaching the film emanate from the top few micrometres only.

8.2. Efficiency in LM autoradiography

8.2.1. Self absorption

The electrons emitted by a disintegrating source in a tissue section are only relevant to the formation of latent images if they can reach the emulsion layer. Figure 23 shows an hypothetical example of the effect of increasing section thickness on accessibility of particles emitted to the emulsion layer. The principle is illustrated graphically in Fig. 24. In situation (a) in Figs 23 and 24 the number of grains formed in a given area of emulsion (and therefore the number of particles arriving at the emulsion) is proportional to the section thickness and the section is said to be 'infinitely thin'. In (b), where the section thickness is increased by an equal increment, not all of the extra particles emitted can reach the emulsion so there is no longer a linear proportionality between source thickness and grain density. In (c), where none of the extra particles reaches the emulsion when the section

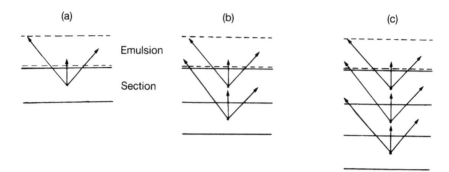

Fig. 23. Vertical section through a generalized LM autoradiograph to show the effect of increasing section thickness on grain density: (a) when all decay particles can reach the emulsion the section is 'infinitely thin' and within this range the grain density is linear with section thickness; (b) when the section thickness is increased only a proportion of the extra decay particles can reach the emulsion so that the grain density is no longer linear with thickness; (c) a further increase in section thickness gives no more decays within the range of the emulsion so that the section is 'infinitely thick'

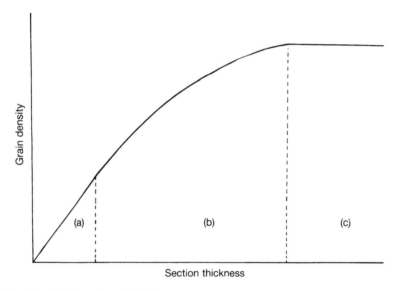

Fig. 24. Graphic illustration of Fig. 23

thickness is increased, there is no increase in grain density and the section is now considered to be 'infinitely thick'.

Various investigators have calculated the percentage penetration of ^{14}C beta-particles in tissue sections within the normal LM range of thickness (i.e. 1–5 μm—see Rogers 1979). Recently, Williams (1987) has measured infinite thickness of ^{14}C polymethyl methacrylate sections with Kodak AR-10. The minimum infinite thickness was shown to be around 85 μm and Williams has therefore recommended that the newly marketed ^{14}C microscales (Amersham International) produced for quantification of LM autoradiographs should be calibrated at a thickness of approximately 100 μm.

Since tritium beta-particles are highly susceptible to self-absorption it would appear that infinite thickness is already reached at 5 μm whereas infinite thinness is somewhat below 0.5 μm (Rogers 1979).

8.2.2. Effect of emulsion on efficiency

It is apparent that increasing emulsion thickness increases grain density where beta-particle range is greater than emulsion thickness. In the case of tritium, the preparation of emulsion layers as described in Chapter 4 will produce thicknesses of 3–4 μm which absorb all beta-particles entering the layer. Hence, all of these emulsion types would be suitable for quantitative work with tritium.

In the case of ^{14}C the long path of beta-particles means that many of these will emerge above the emulsion. Although the density of silver halide can be expected to attenuate the electrons, many micrometres of emulsion would be required to absorb all the beta-particles. In reality, to increase the thickness of the emulsion much beyond 3–4 μm would have such a deterimental effect on resolution that the advantages for quantification and increased efficiency would be negated. Therefore, the practical solution to quantification of ^{14}C autoradiographs at the LM level is to ensure that a uniform emulsion layer is used. For most workers this necessitates the use of AR-10 stripping film. To optimize efficiency of ^{14}C LM autoradiographs an emulsion of high sensitivity such as Kodak NTB-3 or Ilford K5 should be used (Table 3). Conversely, the energy of tritium beta-particles is so low that it is dissipated in low sensitivity emulsions almost as readily as in emulsions of higher sensitivity.

8.2.3. Exposure and efficiency

In general, we expect that prolongation of exposure will increase grain density of autoradiographs. Under certain circumstances, however, this is not so. Where there is fading of latent images (by negative chemography— see Chapter 4) perhaps due to the presence of moisture, the rate of acquisition of silver grains will be less than theoretically predicted.

The other problem of efficiency associated with duration of exposure arises from multiple hits, i.e. the deposition of particle energy in a crystal already containing a latent image. Since the range of tritium particles is relatively short and these hit only crystals close to the source it follows that multiple hits will be a more serious problem with this isotope than with ^{14}C.

8.2.4. Development and efficiency

Anyone who has developed a photographic print knows that there is an optimal point at which the print must be rapidly fixed otherwise it will become totally blackened. The same is true for development of nuclear emulsions in that there is a limit for optimization of signal-to-noise ratio before background grain development (noise) overtakes that of the required latent images. Figure 25 shows the much-quoted data of Caro and Van Tubergen (1962) which illustrate this principle.

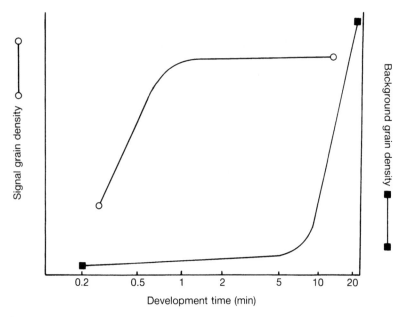

Fig. 25. Development of Ilford L4 at 20°C. The 'signal' grain density reaches a plateau by 2 min (log time scale) whereas the background rises significantly between 5 and 10 min. (Redrawn from Caro and van Tubergen 1962)

8.3. Efficiency in EM autoradiography

Since tritium and iodine-125 are used more than any other radionuclide in EM autoradiography it will be useful to deal with these separately and in greatest detail.

8.3.1. Efficiency of tritium in EM autoradiography

For sections in the usual ultra-thin range (i.e. up to 100 nm) there is only modest self-absorption. For example, self-absorption has been estimated at no greater than 13 per cent in sections 50 nm thick. In addition, staining with uranium or lead salts does not seriously increase self-absorption with tritium and neither does an intermediate carbon layer of 5 nm between section and emulsion.

Vrensen (1970) reported an effect of ultrathin section thickness on ³H efficiency which he attributed to self-absorption of beta-particles although, as already stated, self-absorption should be negligible. Salpeter and Szabo (1972) made further studies with carefully measured section thickness and failed to show a 'percentage efficiency' difference for tritium within the usual ultra-thin range. Table 4 shows a summary of their results. Thus, while doubling section thickness had no effect on efficiency, trebling the dose reduced the efficiency by 20 to 15 per cent. These authors ruled out multiple hits as an explanation for this dose-related effect. They showed, however, that Microdol X increased this dose-dependent loss in efficiency even faster than D19 while gold-Elon ascorbate developer (a latent image intensifying developer) had the least effect in reducing efficiency.

Table 4. *Effect of section thickness (i.e. self-absorption) and radiation dose on sensitivity using ³H methacrylate sections as the source (Ilford L4). (From Salpeter and Szabo 1972)*

Average section thickness (nm)	No. of sections	Average dose (decays/μm²)	Average per cent efficiency
50	8	0.65	20.8
107.5	5	0.61	20.4
94	3	1.8	15

There is little doubt that multiple hits will cause problems where discrete items are heavily labelled or where exposure is unduly long. This phenomenon leads to errors in estimating either absolute or relative radioactivity.

It has been claimed that EM autoradiographs produced by the flat substrate method have a higher efficiency than those prepared with a loop (Salpeter and Salpeter 1971) due to backscattering by the supporting glass slide in the former case. There are no measurements to support this claim. Moreover, if this effect were real it could be expected to have a deleterious effect on resolution since backscattered electrons would subtend a greater solid angle within the emulsion than those entering directly.

At the EM level considerations of efficiency and grain yield as influenced by development are often important. Table 5 gives a useful guide to the relative grain densities which can be expected from tritium by the use of certain combinations of emulsion and developer.

Table 5. *Relative grain densities for Tritium EM autoradiographs produced when certain combinations of emulsion and developer are used. (From Kopriwa 1967)*

Developer	Development time (min)	Development temp (°C)	No. of grains	Relation to D19 (%)	Emulsion
D19b	2	20	175.5	100.00	L4
	2	20	213.0	100.00	NTE
p-Phenylenediamine	1	23	46.6	26.5	L4
	3	23	62.0	35.3	L4
Microdol X	1	20	67.6	38.5	L4
	3	20	76.5	43.5	L4
	3	20	10.0	4.6	NTE

There are few published data on the characteristics of Kodak 129-01 (see Table 3). The manufacturers expect this to be less prone to latent image fading than its predecessor NTE. Doubling the emulsion layer of 129-01 could be expected to increase efficiency much more significantly than in the case of Ilford L4 since far more of the small halide crystals would come within the range of the decay electrons.

8.3.2. Efficiency of iodine-125 in EM autoradiography

Iodine-125 is an isotope which decays by electron capture. Its best known emissions are gamma photons and it is these which are used by biochemists for radiometric counting. Of more importance in EM autoradiography, however, are the Auger electrons most of which are of lower energy than that of the majority of tritium beta-particles. Separation of emulsion from source is more significant for ^{125}I than tritium. Thus sections thicker than 40 nm will undergo self-absorption.

Fertuck and Salpeter (1972) have investigated the sensitivity (efficiency) of ^{125}I in EM autoradiography. These workers examined efficiency as a function of increasing specimen thickness on a semi-log scale. Ilford L4 monolayers were used with gold-Elon ascorbate development. It can be seen from Fig. 26 that increasing section thickness (especially beyond 100 nm) caused a more rapid fall-off in efficiency than was the case with ^3H-labelled sections. A further effect of heavy metal staining on efficiency was shown in this study. Fixed embedded unlabelled albumin, which had been treated with osmium tetroxide and stained *en bloc* with uranyl acetate, was used. Sections of this were interposed between a ^{125}I albumin source (25 nm of fixed embedded ^{125}I albumin) and the emulsion (an L4 monolayer). Table 6 shows that self-absorption was greater where the interposed layer was impregnated with osmium and uranium. As with tritium, doubling of a monolayer of L4 would give little increase in the efficiency of ^{125}I EM

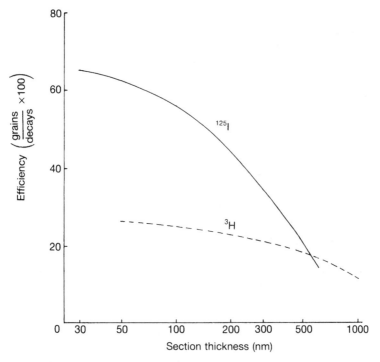

Fig. 26. Decreasing efficiency as a result of increasing section thickness (^{125}I albumin and ^{3}H methacrylate) on a semi-log scale. The soft Auger electrons of the ^{125}I are absorbed more effectively within the specimen than tritium beta-particles. (Redrawn from Fertuck and Salpeter 1974)

Table 6. *Self-absorption with ^{125}I as a function of heavy metal staining. (From Fertuck and Salpeter 1972)*

Thickness of non-radioactive section (nm)	Relative efficiency	
	Epon control	Stained 'tissue'
0	1.00	1.00
25	0.89	0.79
50	0.65	0.59

autoradiographs. On the other hand, there should be a substantial gain in grain density when Kodak 129-01 is changed from a monolayer to a bilayer.

8.3.3. Efficiency of other nuclides in EM autoradiography

It is clear that there is an inverse relationship between the energy of isotope decay particles and efficiency. Harris and Salpeter (1980) have shown that this relationship is not linear. In comparing the different beta-emitters these authors suggest that efficiency is approximately related to the cube root of the energy. For example ^{45}Ca has a 14-fold higher energy than tritium (on the basis of E_{max}) but only 2–3 times poorer efficiency. Similarly, ^{45}Ca has 1.6 times higher energy than ^{14}C but 1.2–1.3 times lower efficiency.

As with ^{14}C and ^{35}S the efficiency can be almost doubled for ^{45}Ca by using a double layer of Ilford L4 emulsion.

8.4. Use of scintillators to improve autoradiographic efficiency

The use of scintillators to enhance the efficiency of autoradiographs has been pursued in various ways during the past two decades. The 'method' has been variously named (e.g. scintigraphy, fluorography, beta-radio-luminescence) and has been the subject of considerable controversy.

Whole body autoradiography has rarely involved attempts to exploit scintillators since the usual combination of ^{14}C radiolabelled compounds and very sensitive X-ray films generally give good autoradiographs after short exposure times.

A specific and useful application of scintillators has been in the macro-autoradiography of ^{3}H-labelled chromatograms and electrophoresis gels. These have very low efficiencies with X-ray films which routinely possess an inert gelatin anti-scratch layer. By incorporating a scintillant in the base material of chromatograms it has been possible to increase efficiencies by a factor of 10 at room temperature. Furthermore, if the temperature of exposure was reduced to $-70°C$ a further 60-fold increase in efficiency was seen. This is because the very small and labile latent images created by light photons (as compared to soft electrons) are stabilized at low temperature. A further factor of 10 was achieved with another manipulation, namely pre-flashing the emulsion layer to low intensity light to build up small sub-latent images in each halide crystal. This helps to reinforce subsequent latent images formed during experimental exposure.

The increases in efficiency claimed for scintillation autoradiography of ^{3}H-labelled specimens at the LM level range from 0 to around 80×. It appears that while it is relatively easy to impregnate specimens with scin-tillators e.g. 2,5-diphenyloxazole (PPO) or 1,4-bis-2-(4-methyl-5-phenyloxazolyl) benzene (dimethyl POPOP), these substances do not readily enter nuclear emulsion layers. Rogers (1981) has conducted a careful review of many of the literature reports. He believes that much of the discrepancy (possibly exaggerated claims for efficiency increase) comes from inadequately controlled baseline efficiency measurements. It seems that many of these are due to insufficiently dried specimens which thus incur latent image fading.

Buchel *et al.* (1978) have claimed increased efficiencies in EM auto-radiography of up to 80 per cent in samples of ³H-labelled T5 bacterio-phages. Layers of the scintillant EPS-9 (50 nm) were applied below, above or on both sides of the Phages. L4 emulsion with gold latensification and Phenidon development was used. It was claimed that increase in efficiency was greatest when there was a layer of scintillator below the specimen. This suggests that a backscattering of photons was necessary to achieve the maximum effect.

It is reasonable to assume that such improvements in efficiency (which are really quite modest) can only be achieved at the expense of resolution. No data have yet been published which quantify the effect on resolution (i.e. image spread). Until we are sure of such effects it is best to avoid the use of scintillators in autoradiography at both the LM and EM levels.

Further reading

Buchel, L. A., Delain, E., and Bouteille, M. (1978). Electron microscope fluoro-autoradiography: improvement in efficiency. *J. Microscopy* **112**, 223–9.

Caro, L. G. and Van Tubergen, R. P. (1962). High resolution autoradiography, 1. Methods. *J. Cell Biol.* **15**, 173–88.

Fertuck, H. C. and Salpeter, M. M. (1980). Sensitivity in electron microscopic auto-radiography for ¹²⁵I. *J. Histochem. Cytochem.* **22**, 80–7.

Harris, W. V. and Salpeter, M. M. (1980). Sensitivity in electron microscopic auto-radiography for Calcium-45. *J. Histochem. Cytochem.* **28**, 40–4.

Kopriwa, B. M. (1967). The influence of development on the number and appearance of silver grains in electron microscopic radioautography. *J. Histochem. Cytochem.* **15**, 501–5.

Longshaw, S. and Fowler, J. S. L. (1978). A poly methyl ¹⁴C methacrylate source for use in whole-body autoradiography and beta-radiography. *Xenobiotica* **8**, 289–95.

Rogers, A. W. (1979). *Techniques of autoradiography* (3rd edn). Elsevier, Amsterdam.

Rogers, A. W. (1981). Scintillation autoradiography at the light microscopic level: a review. *Histochem. J.* **13**, 173–86.

Salpeter, M. M. and Salpeter, E. E. (1971). Resolution in electron microsopic auto-radiography. 11. Carbon-14. *J. Cell Biol.* **50**, 324–32.

Salpeter, M. M. and Szabo, M. (1972). Sensitivity in electron microscopic auto-radiography, 1. The effect of dose. *J. Histochem. Cytochem.* **20**, 425–34.

Williams, M. A. (1987). Infinite thickness in autoradiographs of Carbon-14. *J. Microscopy* **145**, RP1–2.

9 Resolution

In microscopy, 'resolution' means the ability to discriminate spatially between separated points in the specimen. The better the resolution, the smaller the distance between visible points which can be discriminated.

Autoradiographic resolution has been variously defined by a succession of workers and some of these definitions will be described below. Essentially, we are concerned with the relationship between the two-dimensional image spread (array of silver grains) and the radioactive sources within the specimen which give rise to the grains.

Inevitably, a number of factors conspire to ensure that not all of the silver grains produced in an autoradiograph directly overlie their radioactive source. The extent to which this is a problem depends on the level of observation, i.e. macro, LM, or EM autoradiography.

9.1. Resolution in macroautoradiography

On the whole, the main objective in the autoradiography of whole-body sections or chromatograms is to maximize the sensitivity (efficiency) of the emulsion response to small amounts of radiolabel. The final image is one of general blackening rather than discrete grains. Also, the boundaries of the image components are usually easily discernible when viewed with the naked eye such that resolution is not a problem.

However, for purposes of presentation it may be desirable to sharpen the image (increase the contrast) and there are a number of parameters which affect this.

Separation of source and emulsion increases image spread and so the best resolution is obtained with a specimen of minimal thickness and with minimal separation from the emulsion layer.

Higher energy emitters such as ^{32}P and ^{125}I (which produces gamma-photons as well as electrons) give a less sharp image with X-ray film than ^{14}C or ^{35}S. Also, excessive exposure should be avoided since this ultimately leads to latent image formation in crystals distant from the source long after those close to it have been 'saturated'.

9.2. Resolution in light and electron microsoopic autoradiography

One of the most widely used measures of resolution in microscopic auto-radiography is the 'half distance' (HD) which is defined as the distance from a line-source of radiolabel within which one-half of the silver grains are formed. This will be described in more detail below.

If one considers the best spatial resolution which can be achieved with light microscopes, figures of 0.5–2.0 μm will usually be found.

It so happens that most of the HD values for LM autoradiography which have been described are close to the microscopic resolution. This explains why the nature and problems of image spread have been largely ignored by the majority of workers at the LM level.

However, the HD values of the isotopes which have been used at the EM level are approximately three orders of magnitude greater than the best resolution of modern transmission instruments (0.2 nm). Hence silver grains are often formed over cellular compartments different from those containing the source of decay. Although, sad to say, there are still EM autoradiographic studies appearing in the literature which attribute the source of decay to the site of each silver grain, there are no legitimate grounds for this practice. The rest of this chapter deals with various approaches to the problem of characterizing image spread. Of necessity, most of the work done has been at the EM level of autoradiography, but the same principles hold true for light microscopic work.

9.2.1. Factors which determine image spread

When a decay electron (e.g. beta-particle or Auger electron) is emitted, its net trajectory is quite random; in other words it has an equal probability of travelling in any direction. The distance which the electron can travel depends on its initial energy and the density of the surrounding medium (i.e. section and emulsion).

Figure 27 depicts the decay process as we might envisage it in an EM autoradiograph. In reality, fewer than 50 per cent of decay particles have an opportunity to form latent images in the nuclear emulsion. This is because half of them are emitted in a direction away from the emulsion and many of those given off at large acute angles (to the vertical) are stopped within the section before reaching it.

Halide crystals in which stable latent images are formed will, after development, give rise to silver grains. These originate at some unknown point (sensitivity speck) within crystals in which energy is deposited. It is therefore considered that the nature of development has a small but significant effect on image spread.

Thus there is a limited solid cone above the decay source through which electrons contributing to the autoradiographic image can pass. Factors which determine the angle of this cone are section thickness, emulsion thickness, emulsion crystal size, thickness of the intermediate layer, and decay particle energy.

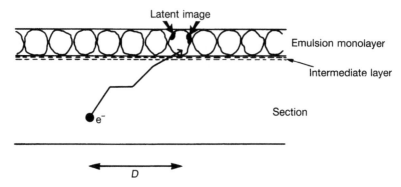

Fig. 27. Schematic vertical section through an EM autoradiograph as we envisage it. The decay electron, e⁻, follows a path in which it is deflected several times until forming a latent image in the emulsion monolayer at a lateral distance D from the point of disintegration

If we consider an imaginary situation, where an infinitely small point source has been allowed to expose beneath a layer of nuclear emulsion to produce an array of silver grains, we might expect to see a grain distribution when viewed from above as shown in Fig. 28a. Here, there are more silver grains per unit area close to the source and the grain density falls off exponentially when plotted against distance from the source (Fig. 28b).

It can be seen that this 'point source density curve' is one way of describing the resolution of the system. Another description of resolution of the same system is via the 'point source frequency curve' which states the probability of a grain forming at a given distance from the source (Fig. 28c). In this case, the maximum probability occurs at a small distance from the source since there are more crystals in close proximity to the source than directly overlying it. This function falls away with distance less steeply than the point source density function.

9.2.2. *Theoretical studies of resolution in EM autoradiography*

In order to understand how present knowledge of resolution has come about it is useful to consider some 'milestone' studies from the past 25 years.

The first important investigation of EM autoradiographic resolution was conducted by Caro (1962) who identified three factors which contribute to the distribution of silver grains about a radioactive source (tritium). These were:

1. The spatial relationship between the source of beta-particles and the crystals hit by the particles.
2. The relation between the passage of a beta-particle through a crystal and the position of the latent image formed.

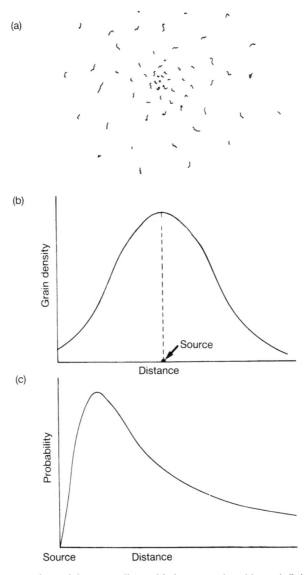

Fig. 28. Representations of the autoradiographic image produced by an infinitely small point source: (a) grains viewed in vertical projection; (b) the corresponding point source density curve; (c) the corresponding point source probability (frequency) curve whose maximum is at a short distance from the source since there are more crystals surrounding the source than immediately above it

3. The relation between the final observed image (i.e. silver grains) and the latent image that produced it. (This concerns the unpredictable growth of silver grains during development which Caro claimed to have overcome by the use of paraphenylene diamine which he erroneously termed a 'physical' developer.)

Caro considered a point source of tritium in methacrylate, 50 nm below an emulsion composed of a single layer of uniform halide crystals 100 nm in diameter. A grain density function was quoted (unfortunately without derivation) as follows:

$$D = D_o \cdot e^{-1.6x}$$

where D is the grain density at distance x and D_o is the grain density over the source.

This predicted image spread function was tested by making autoradiographs of ^{3}H labelled T2 bacteriophages on a collodion film. The distance of each silver grain from its nearest virus was measured. When the above density function was integrated through a sphere of 60 nm diameter (i.e. the diameter of the viral head) the frequency function obtained and plotted agreed well with the measured histogram of grain distances. However, the assumption that each silver grain emanated from its nearest virus particle seems to have reduced the grain frequency at distances greater than 300 nm from the source.

Bachmann and Salpeter (1965) also addressed the problem of image spread about a point source. They considered that there are two error components in respect of the position of a silver grain relative to its decay location. These errors were identified as (a) photographic and (b) geometric (Fig. 29). The mean photographic error was estimated as:

$$E_p = \sqrt{\frac{a^2}{5} + \frac{b^2}{12}}$$

where a is the mean crystal diameter and b is the 'diameter' of the silver grain.

This relationship was used to estimate the photographic error for Ilford L4 and Kodak NTE emulsions (Table 7).

The geometric error was shown to depend on the resolution-limiting thickness (d) of the specimen. For higher energy emitters such as ^{14}C and ^{35}S, scattering was considered negligible and so d was expressed:

$$d = \frac{t_E}{2} + \frac{t_s}{2} + t_i$$

where t_E is the emulsion layer thickness, t_s is the section thickness, and t_i is the thickness of the intermediate (carbon) layer.

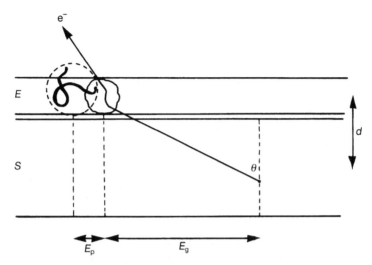

Fig. 29. Schematic vertical section through an EM autoradiograph comprising section, intermediate layer, and emulsion of thickness S, i, and E respectively. The distance from mid-plane of section to mid-plane of emulsion is d. An electron emitted at an angle of θ to the vertical produces a latent image in a crystal at a lateral distance E_g (geometric error). The final developed silver grain is at a distance of $E_g + E_p$ (geometric plus photographic error). (Derived from Bachmann and Salpeter 1965 and Salpeter *et al.* 1969)

Table 7. *Some examples of photographic error. (From Bachmann and Salpeter 1965)*

Emulsion	Developer	Crystal diameter (nm)	Grain diameter (nm)	Mean photographic error (nm)
Ilford L4	Microdol X	100–160	200–300	90
	p-Phenylenediamine	100–160	40–70	60
Kodak NTE (centrifuged)	Dektol	30–55	80–150	40
	Gold latensification/ Dektol	30–55	80–150	40
	Gold latensification/ Elon ascorbate	30–55	40–60	28

When tritium is used, the beta-particles are considered to be significantly scattered which theoretically reduces the thickness of the emulsion available for latent image formation, thus:

$$d = \frac{t_E}{3} + \frac{t_s}{2} + t_i$$

How could this information be used to predict the distribution of silver grains about a radioactive point source? Bachmann and Salpeter resorted to trigonometry to explain this for which we can again refer to Fig. 29.

That fraction of decay electrons which travel within a cone of half-angle

θ to cross the plane of the emulsion $(E) = 1 - \cos \theta$ where the radius of the circle described by the emissions in the plane of the emulsion is $x (= d \tan \theta)$. Figure 30a shows $1 - \cos \theta$ (grain fraction) plotted against $\tan \theta$ (x/d). Thus it can be seen that if the value of d is known, that fraction of the total electrons crossing the emulsion plane at a given distance x can be determined. It can be shown that 50 per cent of all the electrons cross the emulsion within a circle of radius $x = 1.72d$. As θ increases, the value x increases much faster than the number of electrons.

As might be expected, the density of electrons crossing the emulsion plane falls off more rapidly than the fractional count and this is shown in Fig. 30b. Once again, a distinction was made between the assumed non-scattering of ^{14}C and ^{35}S electrons (electron density expressed as $\cos^2 \theta$) and tritium, whose electrons are presumed to be significantly scattered (density expressed as $\cos^3 \theta$).

For the geometric error, E_g, Bachmann and Salpeter, extrapolating from the electron fraction distribution, selected the radius of a circle within which 50 per cent of developed grains fell (this value being equal to $1.72d$ as stated above), i.e. E_g when $1 - \cos \theta = 0.5$. The total error (E_{tot}) was then expressed as:

$$E_{tot} = \sqrt{(E_p)^2 + (E_g)^2}.$$

Calculated values of E_{tot} were published for Kodak NTE and Ilford L4 and were found to be 77 nm and 185 nm respectively. In both cases, the geometric error was considerably greater than the photographic error.

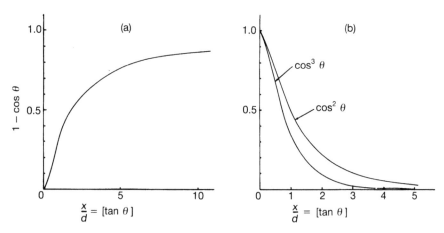

Fig. 30. (a) The fraction of electrons $(1 - \cos \theta)$ which cross the emulsion emitted from a point source within a cone of half angle θ plotted against radius of radiation field in the emulsion x/d $(x = E_g$ in Fig. 29). (b) The density of developed grains plotted against x/d (tan θ). In the case of tritium, $\cos^3 \theta$ is more appropriate representing scattering of beta-particles and therefore lower grain densities. The curve of $\cos^2 \theta$ represents ^{14}C and ^{35}S where negligible scattering occurs. (Redrawn from Bachmann and Salpeter 1965)

9.2.3. Measurements of resolution (image spread) in EM autoradiography—line sources

Although the above mathematical models aided the understanding of image spread, in practice they did not have any great impact on the routine assessment of autoradiographic images.

A key study, which was to have a major influence on the interpretation of EM autoradiographs in particular, was reported in a rather complicated paper by Salpeter *et al.* (1969). In this new approach, a very thin line source of tritium was created. This was achieved by dissolving ³H-styrene in an organic solvent from which films were formed on glass slides (as ³H-polystyrene). These films were then cast on to a water surface, picked up on to a flat Epon block, and overlayed with a 3 μm methacrylate film. When this 'sandwich' was sectioned at right angles, the radiolabelled polystyrene appeared as a thin uniform line about 50 nm thick. EM autoradiographs were then made from these sections containing the line source using Ilford L4 (about 140 nm) or Kodak NTE (about 70 nm) monolayers. The Ilford emulsion was developed with either Microdol X or paraphenylene diamine and the Kodak NTE with either Dektol (= Kodak D-163) or gold latensification/elon ascorbate.

It was seen in the final autoradiographs that not all the silver grains were formed directly above the line which was known to be the source of all the radioactive decays. Figure 31 shows an example of such a 'hot-line' EM autoradiograph from the author's own work with chromium-51.

A grain density distribution from the tritium line source was obtained by

Fig. 31. Radioactive line source of chromium-51—EM autoradiograph. Note that the silver grains are produced at varying distances from the source. Scale bar = 1 μm

measuring the distance from the mid-point of each grain to the nearest point on the source line. A grain number historgram from such a line source is shown in Fig. 32 and a 'best fit' curve for this distribution can be seen in Fig. 33. Salpeter and her colleagues found that the spread of the grains was greater for Ilford L4 than Kodak NTE as would be expected from the larger crystal diameter of the Ilford emulsion.

A convenient way to express image spread about a line source is as a cumulative (or integrated) curve in which grains are added consecutively with increasing distance from the source (Fig. 34). As mentioned at the beginning of this chapter, the distance within which half of the silver grains produced from the line source fall is known as the 'half-distance' (HD) and this value must be defined for isotope, section thickness, emulsion type, emulsion thickness, and developer. The value HD has, for many workers, become synonymous with resolution of the autoradiographic system. It must be remembered, however, that the HD value defines only one point on the grain density curve and strictly, therefore, it is the latter which fully

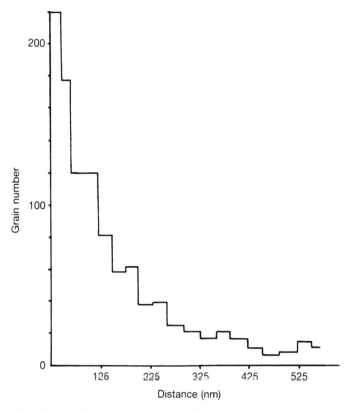

Fig. 32. Density histogram from a line source of tritium. EM autoradiograph prepared with grey sections, Kodak NTE, and developed in Dektol. (Redrawn from Salpeter *et al.* 1969)

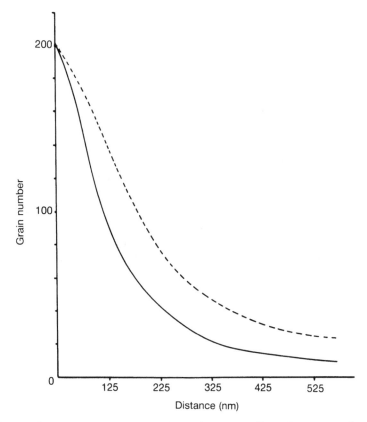

Fig. 33. Best fit curves for grain density about a line source. The solid curve was fitted to the histogram in Fig. 32 while the dashed curve shows grain spread resulting from the same sections after exposure to Ilford L4 and development in Microdol X. (Redrawn from Salpeter *et al.* 1969)

describes the resolution. Nevertheless, HD is a useful reference when comparing the resolution of different systems.

In their 1969 paper, Salpeter *et al.* introduced the further concept of 'universal curves'. These are grain density curves with the grain distance (from source) values normalized to units of HD. Universal curves and their use will be described further in Chapter 10.

9.2.4. Measurements of resolution (image spread) in EM autoradiographs—point sources

As described above, the distances of the mid-point of silver grains from the line source are measured as the distance from the closest point on the line (i.e. the perpendicular distance from the line). In reality, we have no way of knowing where on this line is the point of decay from which each grain arises and it is therefore probable that the closest point on the line source to every grain will not be the origin of its associated decay particle.

Hence, in terms of point sources (which is what we are normally concerned with), the curves in Figs 33 and 34 represent underestimates of image spread. Figure 35 shows cumulative grain density plots for both line and point sources of chromium-51. It can be seen that the spread of grains about a point source is greater than that about a line source. Moreover, we now have to think in terms of 'half-radius' (HR) i.e. the radius of a circle within which one-half of the grains produced by a point source can be expected to occur. For a given isotope and under identical conditions, HR = 1.73 HD.

How are the 'best fit' curves for line and point source grain distributions derived? For details of the mathematics the reader is referred to the appendix of Salpeter *et al.* (1969) and to Blackett *et al.* (1980).

Briefly, the basic functions are as follows. The cumulative theoretical line source curve was derived:

(i)
$$P_1 = \frac{2}{\pi} \tan^{-1} \frac{x}{d}$$
(Salpeter *et al.* 1969)

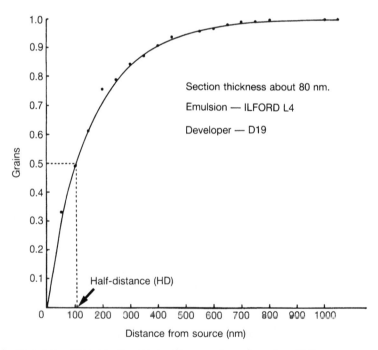

Fig. 34. Cumulative (integrated) grain density curve for a chromium-51 line source prepared under the conditions shown. The 'half-distance' (HD) is 108 nm

where P_1 is the cumulative probability, x is the grain distance from the line source in HD units, and d is the vertical distance from the source to the mid plane of the emulsion (Fig. 29).

When equation (i) was fitted to the measured cumulative line source curve a reasonably good fit was achieved, although the tail of the curve did not fit very well. The function has been modified to give a better fit for the tail:

(ii)
$$P_1 = \frac{2a \tan^{-1} \frac{x}{d}}{\pi} + \frac{bx}{\sqrt{c^2 + x^2}}$$
(Blackett *et al.* 1980)

where a, b, c, d are variables for curve fitting detailed in Blackett *et al.* (1980) for a number of beta and Auger emitting isotopes. Having used the experimentally measured line source data to refine the theoretical function for a line source, it was possible to derive an optimal point source probability function by considering a line as a series of point sources:

(iii)
$$p = \frac{adx}{(x^2 + d^2)^{1.5}} + \frac{2bc^2 x}{(c^2 + x^2)^2}$$
(Blackett *et al.* 1980)

where p is the probability that a grain is produced at a distance x (in HD units) from the point source.

Equation (iii) for point source frequency distribution (or equation 5b in

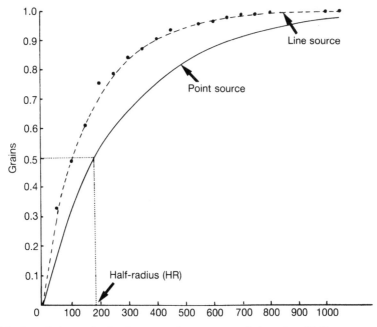

Fig. 35. Cumulative grain density curves for a measured chromium-51 line source and its derived point source density curve for which the 'half-radius' (HR) is 187 nm

the mathematical appendix of Salpeter *et al.* 1969) forms the basis for the most refined methods of analysis currently used in EM autoradiography (see Chapter 10).

Further reading

Bachmann, L. and Salpeter, M. M. (1965). Autoradiography with the electron microscope. A quantitative evaluation. *Lab Invest.* **14**, 303–15.

Blackett, N. M., Parry, D. M., and Baker, J. R. J. (1980). Isotope decay range distribution curves for use in the analysis of electron microscope autoradiographs. *J. Histochem. Cytochem.* **28**, 1050–4.

Caro, L. G. (1962). High resolution autoradoigraphy. 11. The problem of resolution. *J. Cell Biol.* **15**, 189–99.

Salpeter, M. M., Bachmann, L., and Salpeter, E. E. (1969). Resolution in electron microscope radioautography. *J. Cell Biol.* **41**, 1–20.

10 Quantification of EM autoradiographs

In Chapter 9 we considered the nature of image spread, in other words the mathematical certainty that not all silver grains overlie the decay sources which give rise to them. The consequence of this is that, in EM autoradiography, some silver grains are produced over structures other than those which contain the causal radiolabel. This phenomenon is known as 'cross-fire'. It is the purpose of this chapter to indicate how our knowledge of image spread can be used to correct for cross-fire in experimental situations.

Williams (1982) has coined the terms 'restricted' and 'unrestricted' to describe two fundamental choices for analysing autoradiographs. Restricted methods are suitable when there are grounds for supposing that the autoradiographic image arises from discrete radiolabelled sources of simple and regular geometry (circles, discs, bands, etc.). The aim is then to compare known grain distributions about regular model sources with those about regular (or near regular) sources within the specimen.

More often, the grain distribution over the tissue sections offers few, if any, clues about the pattern of labelling. In this case an unrestricted method of analysis is needed. This involves unbiased morphometric measurement of tissue organization plus the definition of locations of grains. This process must take into account the capacity for generating cross-fire of each tissue component. Thence predicted intensities of labelling for each feature can be compared with observed grain densities of these features and corrected estimates of activity within each of the sources derived.

10.1. Restricted methods

10.1.1. Analysis of structures of similar shape and size

Salpeter *et al.* (1969) noted that if the grain density curves derived from tritium line sources of different resolution were plotted with distance from the source on the x-axis expressed in terms of units of HD they always appeared to have the same shape. These curves were named 'universal curves' (Section 9.2.3). Apart from grain density being normalized to 1 at the origin and distance expressed in HD units the universal curve shape for a line source is as shown in Fig. 33.

Since, however, few radiolabelled sources in biological specimens present themselves as straight lines, it is necessary to use the mathematical description of image spread to predict grain distributions about labelled structures of regular geometry. Basically, point source density functions (see mathematical appendix of Salpeter *et al.* 1969) are integrated to obtain functions for discoid, annular, and band-shaped sources and some of the corresponding curves are presented in Figs 36 and 37.

Provided that the HD of the system is known (for tritium, HD is 165 nm for a 120 nm section, Ilford L4 monolayer, and Microdol X development) the universal curves can be used to test hypotheses of label distribution in structures of similar and regular shape. It is also necessary that the structures should be of similar size.

The experimental grain distribution can be obtained by constructing histograms of the number of grain centres per unit area both inside and outside the boundary of the labelled structures (e.g. discs). Once all the data are collected the grain density histogram is normalized. This is done by dividing the *x*-axis values by HD and the *y*-axis values by the density at the boundary.

For comparison with the normalized experimental density histogram, the appropriate model normalized universal density curve is selected. This

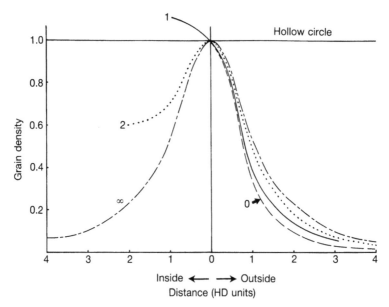

Fig. 36. Grain density universal curves for hollow circular sources (annuli) of radius (in HD units) 0, 1, 2, and infinity (∞). In the extreme case where HD = 0 and infinity the sources are in reality a point and a straight line respectively. The density value at the circumference is normalized to unity. Hence when HD is small (e.g. 1) the grain density inside the circle is greater than at the source (circumference). (Redrawn from Salpeter *et al.* 1969)

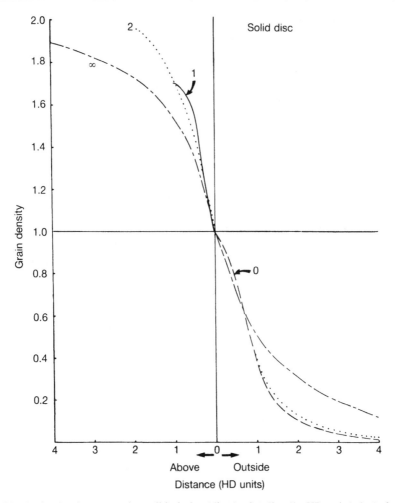

Fig. 37. Grain density curves for solid circles (discs) of radius (in HD units) 0, 1, 2, and infinity (∞) normalized to unity at the outer edge. Once again zero HD is equivalent to a point source. (Redrawn from Salpeter *et al.* 1969)

means for the half dozen curves derived by Salpeter *et al.* for each source shape, one must be chosen according to its radius (if circular or annular) or 'half-width' (if a band) such that the theoretical curve nearest the dimensions of the experimental structures is used. A more convenient way to compare experimental with theoretical results is to use the integrated (cumulative) curves (e.g. see Fig. 34) such that the proportion of grains formed at a given distance from the origin (i.e. boundary) may be compared. In other words it is possible to see what percentage of grains is expected to fall over a radioactive source and what percentage outside it.

In this restricted method the initial assumption is that the source is uniformly labelled. It is possible, by trial and error, to generate curves representing superimposed sources such as an annulus over a disc of identical outer diameter. Such an approach was used to explain the distribution of silver grains about transverse sections of ^{3}H-noradrenaline labelled sympathetic nerve terminals (Budd and Salpeter 1969) where 86 per cent of label could be ascribed to the whole terminal but the remaining 14 per cent was concentrated around the periphery.

10.1.2. Analysis of structures of similar shape but dissimilar size

It should be understood that the method outlined in Section 10.1.1 has several limitations. The most important of these is that two-dimensional structures only of very similar size as well as shape can be assessed by reference to a given universal curve (Downs and Williams 1984).

In practice, electron microscopists are more often concerned with sections of spherical structures (vesicles, granules, droplets, some mitochondria) which thus contain discs or annuli having a wide range of radii but which must be analysed as one system.

The paper of Downs and Williams (1984) offers detailed derivations of composite probability functions for uniformly labelled discs, annuli, and rectangles of a range of sizes. From these functions, tables are provided which show percentages of grains which may be expected over and outside a series of multisized discs, squares, and rectangles. Dimensions once again are expressed in units of HD.

A brief account of this very complex mathematical paper and its use in autoradiographic analysis will be given here. Indeed, despite the alarming number of functions which appear in the text and appendices, the tables in the second appendix contain vast amounts of data on image spread which are readily usable (see Tables 8 and 9).

10.1.2.1. Combined and cumulative grain frequency functions

The image spread from a population of radiolabelled discs is first described.

The grain frequency function is defined for a set of n discs, uniformly labelled and superimposed concentrically (as envisaged for profiles of a spherical granule sectioned serially). If the ith disc is assumed to have a radius of C_i (measured in HD units) and radiolabel concentration K_i

Table 8. *Uniformly-labelled disc-shaped sources; theoretical percentages of grains (F_D(C, R) × 100) within concentric circular regions, based on idealized frequency function for discs of radius 1 to 10 HD units. (In HD units, C = radius of disc, (C + Δ) = radius of circular region; PS = point source.)*

Δ	PS	1	2	3	4	5	6	7	8	9	10
−9											0.9
−8										1.1	3.6
−7									1.4	4.4	8.1
−6								1.8	5.5	9.8	14.3
−5							2.3	7.0	12.2	17.4	22.2
−4						3.2	9.2	15.6	21.6	27.0	31.8
−3					4.7	12.7	20.5	27.4	33.3	38.4	42.8
−2				7.5	18.4	27.8	35.6	41.9	47.1	51.4	55.1
−1			13.1	28.1	39.0	47.1	53.2	58.0	61.9	65.0	67.7
0		23.2	42.2	53.3	60.5	65.6	69.5	72.4	74.8	76.8	78.5
1	29.3	52.4	63.1	69.4	73.6	76.6	78.9	80.8	82.3	83.5	84.6
2	55.3	67.2	73.4	77.3	80.0	82.1	83.7	85.0	86.0	86.9	87.7
3	68.4	75.2	79.2	81.8	83.8	85.3	86.5	87.4	88.2	88.9	89.6
4	75.7	80.1	82.8	84.8	86.3	87.4	88.3	89.1	89.8	90.3	90.8
5	80.4	83.4	85.4	86.9	88.0	89.0	89.7	90.4	90.9	91.4	91.8
6	83.6	85.8	87.3	88.5	89.4	90.2	90.8	91.3	91.8	92.2	92.5
7	85.9	87.6	88.7	89.7	90.5	91.1	91.6	92.1	92.5	92.8	93.2
8	87.6	89.0	89.9	90.7	91.3	91.9	92.3	92.7	93.1	93.4	93.7
9	89.0	90.1	90.8	91.5	92.1	92.5	92.9	93.3	93.6	93.8	94.1
10	90.0	91.0	91.6	92.2	92.7	93.1	93.4	93.7	94.0	94.2	94.5
12	91.7	92.4	92.8	93.3	93.6	94.0	94.2	94.5	94.7	94.9	95.1
14	92.9	93.4	93.7	94.1	94.4	94.6	94.8	95.0	95.2	95.4	95.5
16	93.8	94.2	94.4	94.7	94.9	95.2	95.3	95.5	95.7	95.8	95.9
18	94.5	94.8	95.0	95.2	95.4	95.6	95.8	95.9	96.0	96.1	96.3
20	95.0	95.3	95.4	95.6	95.8	96.0	96.1	96.2	96.3	96.4	96.5

Reproduced from Downs and Williams (1984) by permission of the authors and *The Journal of Microscopy*

($i = 1, \ldots, N$), for any distance R (in HD units) measured from the common centre of the discs, the combined frequency function $f_D^*(R)$ is given by:

$$f_D^*(R) = \frac{\left(\sum_{i=1}^{N} K_i C_i^2 f_D^*(C_i, R) \right)}{\left(\sum_{i=1}^{N} K_i C_i^2 \right)} \tag{i}$$

From this a cumulative frequency function $F_D(R)$ would be expressed:

$$F_D(R) = \int_0^R f_D^*(R) \, dR \tag{ii}$$

Table 9. *Uniformly-labelled disc-shaped sources; theoretical percentages of grains ($F_D(C, R) \times 100$) within concentric circular regions, based on idealized frequency function for discs of radius 12 to 50 HD units. (In HD units, C = radius of disc, (C + Δ) = radius of circular region.)*

Δ	C										
	12	14	16	18	20	25	30	35	40	45	50
−45											1.0
−40										1.2	3.9
−35									1.5	4.8	8.8
−30								2.0	6.1	10.9	15.7
−25							2.7	7.9	13.7	19.3	24.4
−20						3.8	10.7	17.8	24.3	30.1	35.1
−18					1.0	7.5	15.4	22.9	29.4	35.0	39.9
−16				1.2	3.8	12.4	21.0	28.5	34.9	40.4	45.0
−14			1.5	4.7	8.5	18.5	27.4	34.8	40.9	46.1	50.5
−12		1.9	5.9	10.5	15.2	25.8	34.6	41.7	47.4	52.1	56.2
−10	2.6	7.6	13.1	18.6	23.6	34.3	42.6	49.1	54.3	58.6	62.2
−8	10.2	17.0	23.3	28.9	33.9	43.9	51.4	57.1	61.6	65.3	68.5
−6	22.7	30.0	36.2	41.4	45.9	54.6	60.9	65.7	69.4	72.4	75.0
−4	39.9	46.4	51.6	55.9	59.4	66.3	71.0	74.6	77.4	79.7	81.7
−2	60.9	65.4	68.9	71.7	74.0	78.4	81.4	83.7	85.5	87.1	88.4
−1	71.8	75.0	77.5	79.5	81.2	84.2	86.4	88.0	89.3	90.5	91.6
0	81.1	83.1	84.7	85.9	87.0	89.0	90.4	91.5	92.4	93.3	94.1
1	86.3	87.6	88.7	89.6	90.3	91.6	92.6	93.4	94.1	94.8	95.5
2	89.0	90.0	90.8	91.4	92.0	93.0	93.8	94.4	95.0	95.6	96.2
4	91.7	92.3	92.9	93.3	93.8	94.5	95.0	95.5	95.9	96.4	97.0
6	93.2	93.7	94.1	94.4	94.7	95.3	95.7	96.1	96.5	96.9	97.4
8	94.1	94.5	94.9	95.1	95.4	95.8	96.2	96.5	96.9	97.3	97.7
10	94.8	95.2	95.4	95.7	95.9	96.2	96.5	96.8	97.1	97.5	97.9
12	95.4	95.6	95.9	96.1	96.2	96.6	96.8	97.1	97.4	97.7	98.1
14	95.8	96.0	96.2	96.4	96.6	96.8	97.0	97.3	97.5	97.9	98.3
16	96.2	96.4	96.5	96.7	96.8	97.0	97.2	97.4	97.7	98.0	98.4
18	96.5	96.6	96.8	96.9	97.0	97.2	97.4	97.6	97.8	98.1	98.5
20	96.7	96.9	97.0	97.1	97.2	97.4	97.5	97.7	97.9	98.2	98.6

Reproduced from Downs and Williams (1984) by permission of the authors and *The Journal of Microscopy*

10.1.2.2. Application to analysis of autoradiographs

If the sources of radiolabel in the sections are approximately circular and sufficiently separated so that their radiation fields do not overlap, it is possible to test the hypothesis that the grain distribution is consistant with uniform labelling of the sources.

Source profiles (discs) are selected or excluded from a representative set of micrographs as described in Fig. 38. Cut-off distances may be defined in terms of the percentage of grains associated with a labelled source which one wishes to remain non-overlapped by the radiation field of an adjacent structure. Tables 8 and 9 (Downs and Williams' A1a and b) show that for

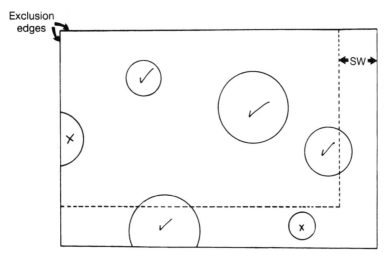

Fig. 38. Objective method for deciding which circular profiles are to be included in an analysis. Profiles are excluded when they intersect the two exclusion edges or lie completely outside the inclusion edges (broken lines). The distance of the inclusion edges from the print boundary, the strip width (SW), is determined such that the centres of all included profiles lie within the full field. (Redrawn from Downs and Williams 1984)

90 per cent of source-associated grains a disc of radius 2 HD requires a cut-off distance of just over 8 HD while for a disc of radius 10 HD, a distance of a little more than 3 HD will do.

When cut-off distances have been determined, overlapping and incomplete regions are identified as depicted in Fig. 39 using transparent overlay screens with circles of different HD radii designed to give the desired percentage grain association per profile (e.g. 90 per cent). Thus if alpha i is the total excluded angle in degrees for the ith disc, the fraction of the ith disc width included in the analysis is expressed as:

$$w_i = 1 - \frac{a_i}{360}$$

The radii of sources and any excluded angles are measured and recorded together with any excluded sector angles. Sets of concentric circles on an overlay screen are used to record silver grains which are categorized (in histogram bins) according to radial distance from the profile centre (Fig. 40).

If there are j histogram classes the expected number of grains in the jth class (E_j) is given by:

$$E_j = [N_G F_D(R_j) - F_D(R_{j-1})]$$

where N_G is the total number of grains recorded and $j = 1, \ldots, n$. The cumulative frequency function values, $F_D(R_j)$, can be calculated using

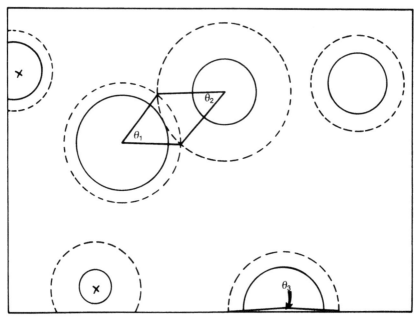

Fig. 39. Data collection excludes regions which overlap and is determined by cut-off circles (dotted) which include at least 90% of grains from a profile and whose radius depends on the size of the source profile. The angles of sectors to be excluded from the field shown are θ_1, θ_2, and θ_3. (Redrawn from Downs and Williams 1984)

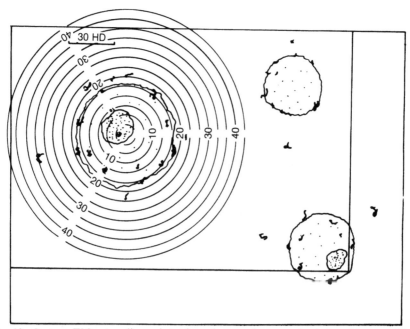

Fig. 40. Cartoon EM autoradiograph to show the use of concentric circles of known HD to collect grain data. (Redrawn from Downs and Williams 1984)

equations (ii) and (i) in Section 10.1.2.1. For profiles with excluded sectors (e.g. disc *i*) the area weighting factors (C_i-squared) must be multiplied by w_i (described above) to compensate for the expected proportional reduction of grains.

The expected and recorded grain values are compared using the chi-squared test. A poor fit necessitates revision of the expected values. For example, the assumption that the discs are uniformly labelled may not be valid and it may be necessary to treat each profile as a combination of discs and annuli with varying densities of radiolabel.

Table 8 shows theoretical grain percentages expected within given distances (in HD units) from the centre of uniformly labelled discoid sources of radius 0 to 10 HD. Table 9 shows similar data for circle radii up to 50 HD. If a fully comparative analysis as described above is not required, these tables can be used to assess whether individual disc-shaped profiles conform to the prediction of uniform labelling.

10.2. Unrestricted methods

10.2.1. The 'circle' method

This method was devised by Williams (1969) to allow some definition of the location of silver grains with respect to the ultrastructure of the specimen. Basically, it tests the hypothesis that the radiolabel is randomly distributed (or not) and ascribes values for the relative concentration of radiolabel within given structures. Moreover, in so doing the method takes into account part of the cross-fire (nominally 50 per cent). Although this lacks the precision of later methods, it has the advantage of simplicity (no computer is needed) and is capable of giving useful comparative results.

10.2.1.1. Effective area measurement

Firstly, it is necessary to ensure that a set of EM autoradiographs has been collected which constitutes a fully representative sample of the tissue or cells under study. Let us suppose that the simple cartoon in Fig. 41 represents an EM autoradiograph of our experimental cell type and that we have a pile of similar pictures. It can be seen that, excluding boundaries, the sample can be divided into 'primary items' namely cytosol, nucleus, granules, mitochondria, and extracellular space.

The unbiased sampling of points on sufficient pictures of a population gives a measure of the relative area of the primary items (equivalent to volume fraction in three dimensions). If an adequately large number of pictures is sampled the same result can be achieved by regular (e.g. quadratic) sampling of the items. However, sampling at infinitely small points gives little or no information about the juxtaposition of structures

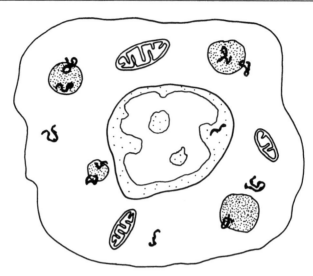

Fig. 41. Cartoon EM autoradiograph of a single cell in which the 'primary' items can be designated as nucleus, mitochondria, granules, cytosol, and extracellular space

and since the latter may influence patterns of labelling there is merit in sampling finite areas rather than points.

Figure 42 illustrates the way in which effective (relative) area is in fact measured. A quadratic array of circles in the form of a transparent overlay is fixed in a consistant way to the series of micrographs. The radius of the circles is conventionally (and conveniently) taken as the half-radius (HR) of the system i.e. the circles are 50 per cent probability circles. Table 10 lists HR values for tritium under various experimental conditions.

The structures in each circle are recorded systematically and the items categorized and totalled as in the first column of Table 11. Sometimes the structures recorded are single or 'primary' items such as cytosol or nucleus. More often, however, they are 'junctional' (e.g. cytosol/extracellular) or 'compound' (e.g. cytosol/mitochondrion/extracellular) items and Table 11 lists the permutations which might be expected from circle sampling of cells such as in Fig. 42.

Table 10. *HR (50% probability circles) for tritium derived from HD values given by Salpeter et al. (1969) (HR = 1.73 × HD)*

Section thickness (nm)	Emulsion	Developer	Half-radius (HR) (nm)
120	L4	Microdol X	280
120	L4	Paraphenylene diamine	245
50	L4	Microdol X	245
50	L4	Paraphenylene diamine	220
120	NTE	Dektol	170
50	NTE	Dektol	135

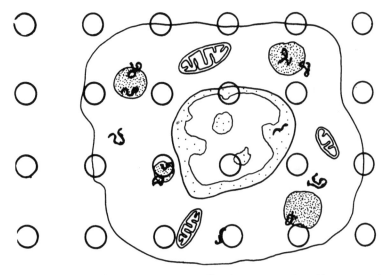

Fig. 42. Relative area can be measured by recording the contents of a uniform array of circles which in this case are of radius HR, i.e. 50% probability circles. In this way categories are created which contain single and multiple items

In real situations the list of items can be very long and so it becomes necessary to group together some small totals from circles containing structures in common. Examples of this procedure appear in column 2 of Table 11. We now have a regular (equivalent to random), unbiased sampling of the relative area of the cell components.

10.2.1.2. Recording of silver grains

Now we can turn our attention to the actual silver grains. Fifty per cent probability circles are placed symmetrically around every silver grain in the set of autoradiographs (i.e. the centre of the longest axis of the grain is at the centre of the circle) and the structures underlying the circle are categorized and totalled as before (see column 3 of Table 11). These are the observed values (O).

10.2.1.3. Manipulation of circle and grain data

Since the method uses chi-squared to test the randomness (or otherwise) of the radiolabelling it is necessary to know how the number of real grains present would be located if they were randomly distributed. These expected grain values ((E) in column 4 of Table 11) are obtained by multiplying the grain total by the appropriate fraction of total circles recorded for a given

Table 11. *Data summary and calculation for the 'circle' method*

Item	No. of circles	No. of circles after grouping	No. of grains (O)	No. of grains if random (E)	$\chi^2 = \dfrac{(O-E)^2}{E}$	Relative specific radioactivity (grains/circles)
Cytosol	400	400	207	233.0	2.90	0.518
Cytosol/Mitochondria	98	98	28	57.1	14.75	0.286
Cytosol/Extracellular	123 ⎱	224	100	130.4	6.92	0.446
Cytosol/Extracellular Mitochondria	101 ⎰					
Cytosol/Granule	391 ⎱	640	698	372.8	283.18	1.091
Cytosol/Granule/ Extracellular	249 ⎰					
Nucleus	295 ⎱	509	102	296.5	127.15	0.200
Cytosol/Nucleus	214 ⎰					
Cytosol/Nucleus Granule	148	148	41	86.2	23.55	0.277
Total	2019	2019	1176	1176	458.45 $p < 0.001$ (5d of f)	

item. Thus the expected number of grains (E) over cytosol from a random distribution would be:

$$(E) \text{ cytosol} = 1176 \times 400/2019 = 233$$

and so on.

Before the individual chi-squared values (as in column 5 of Table 11) are calculated, expected (E) values of 5 or less must be either ignored in calculating chi-squared as they are too unreliable or eliminated by further grouping.

The larger the individual chi-squared value for each item, the greater is its deviation from randomness with respect to its radiolabel content. Since there are 6 grouped items, 5 degrees of freedom are allowed for the chi-squared test. Reference to tables of chi-squared show that for 5 degrees of freedom the total of 460.96 in Table 11 gives a *p* value of < 0.001. This means that there is less than 1 in 1000 chances that the silver grain distribution has arisen from a random distribution of radiolabel.

Finally, in order to obtain estimates of the relative concentration of label (relative specific radioactivity) in each item, the number of grains for that item is divided by the number of circles. This is equivalent to dividing the proportion of grains by the proportion of circles since each is based on the same totals, 1176 and 2019 respectively, which would cancel out in division. It can be seen from these activity estimates in column 6 of Table 11 that the grouped item in which granules are the common feature contains the most concentrated radiolabel in this make-believe experiment.

Since we have used 50 per cent probability circles to describe grain positions we can suppose that for a sufficiently rigorous sample we have taken approximately the first 50 per cent of image spread into account using this method.

10.2.2. The 'hypothetical grain' method

The 'universal curves' of Salpeter describe full image spread from sources of restricted shape (and size) and the 'circle' method of Williams allows partial image spread assessment for any source.

The quantum leap which permitted the evaluation of the full range of image spread (and thus 'cross-fire') for all sources in any specimen was made by Dr N. M. Blackett (Blackett and Parry 1973, and a more refined account in 1977). The system devised by Blackett is known as the 'hypothetical grain' method.

An outline description of the method will be given here although the reader is referred to Blackett and Parry (1977) for more detail. The procedure can be divided into 5 stages: (i) preparation of an overlay screen which represents the image spread of the system; (ii) collection of 'hypothetical grains' and construction of a cross-fire matrix; (iii) collection of real grain data; (iv) 'fitting' of hypothetical grain data to real grain data using the chi-squared test; and (v) if necessary, modifying the matrix until an acceptable fit is obtained.

10.2.2.1. Construction of an overlay screen used for cross-fire measurement

The method centres on the need to know the full extent of image spread of the system. Therefore, a set of distances representing complete image spread within the specimen must be established. In fact, these cross-fire distances are found by random sampling of the *y*-axis of the appropriate point source probability curve (e.g. Fig. 28c and Section 9.2.4, equation iii) and reading off the corresponding distances on the *x*-axis.

Let us now test the hypothesis that our specimen is uniformly labelled rather as we did with the 'circle' method by recording the position of regular (or random) circles. This time however, for greater precision we want to ascribe activity to discrete points rather than associations of items. One way of doing this is to place an overlay screen containing a quadratic array of intersections and recording the single structures which appear directly below these 'grid points'.

These structures are then denoted 'sources' and are the points from which, for the sake of our hypothesis, we are supposing disintegrations to emanate. To recap, we have created an hypothesis that there is one decay per unit area in the specimen and to illustrate the process so far we can refer to the cartoon autoradiograph in Fig. 43 in which the sources of decay are indicated by the intersections.

Having defined the origins of these decays, we now need to identify the positions ('sites') of the hypothetical grains which the decays might be expected to produce in the emulsion. The distances used are those discussed above and chosen sequentially from the random sampling of the point source probability curve. The centre of each hypothetical grain is defined by the centre of a circle of radius HR. Since the initial trajectory of the decay electrons is random, the putative direction of each hypothetical grain is also randomly chosen. The situation described thus far is seen in Fig. 44 which shows a simple screen with decay points and their associated hypothetical grains (circle centres).

Circles (which happen to be 50 per cent probability circles) are used since they conveniently enclose the finite size of the coiled silver grains and account for the frequent overlap by grains of more than one structure (remember, however, that the decay which gives rise to each grain is envisaged as an infinitely small point). Furthermore, there are statistical advantages in using circles to represent hypothetical grains which will be

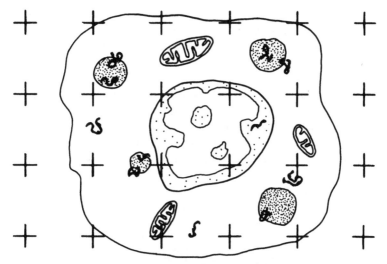

Fig. 43. Representation of an hypothesis of uniform labelling. This is equivalent to the assumption that the decays are produced solely at the points of intersection of the grid

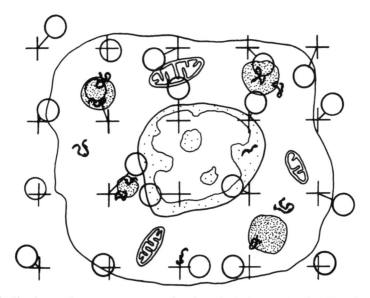

Fig. 44. Simple overlay screen representing hypothetical sources of radioactive decay (uniform intersections) and the sites of their associated hypothetical grains (HR circles—the centre of each hypothetical grain is the centre of the circle). The distances between sources and sites of hypothetical grains are determined by random sampling of the point source probability curve for the system. The directions of decay are randomly chosen

discussed later (Section 10.2.2.4). In practice, the overlay screen used is shown in Fig. 45. Whilst alarming at first sight, this screen is comparable to that in Fig. 44 but now contains 390 hypothetical decay-to-grain (source-to-site) distances. This represents an adequate sampling of the point source probability curve. This full screen differs from that in Fig. 44 because, instead of the uniform quadratic sources of decay, it is the hypothetical grains (circles) which, for convenience, are now in uniform hexagonal array. There are 10 distances (numbered 0 to 9) associated with the centre of each circle. Each isotope has its own characteristic screen whose magnification must of course be scaled to that of the prints used for analysis.

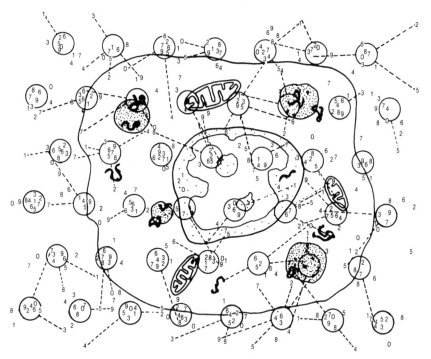

Fig. 45. The full and practical version of the overlay screen shown in Fig. 44. Instead of sources being uniformly positioned the sites are now in uniform hexagonal array. Sources are represented not by intersections but by numbers from 0 to 9 associated with each circle. Many more hypothetical decays are therefore included on one screen

10.2.2.2. Collection of hypothetical grain data

There are several ways in which the screen can be used to generate an hypothesis of labelling. Bearing in mind our intention to test initially the hypothesis of uniform labelling, the following method is recommended.

To each of a large and representative set of prints the screen is pinned in a consistant way and the '0' point sources are chosen first. Taking each circle in turn the single source (of decay) item beneath each '0' is noted as are the contents of the associated circle (hypothetical grain) and a single score made in the appropriate box of a source-to-site matrix. For the next print all the number 1's are chosen and so on until sufficient hypothetical grains (usually at least 1000) have been recorded. Table 12 shows how such a 'cross-fire matrix' might look if our fictional specimen in Figs 41–45 had been so recorded.

For large sources, the number of hypothetical grains produced is greatest over the same item. In addition, some grains might be expected to fall on adjacent items. For example, decays from the cytosol have produced 18 hypothetical grains over cytosol/mitochondria which constitutes part of the cross-fire of the system. The final column in Table 12 reminds us that we have collected hypothetical grain data consistant with the hypothesis of uniform labelling.

10.2.2.3. Collection of real grain data

The actual silver grains are recorded by placing a 50 per cent probability circle symmetrically around each grain in turn and scoring its position according to the site categories in the columns of the cross-fire matrix. Some imaginary real grain totals are shown in the bottom row of Table 12.

10.2.2.4. Testing the hypothesis

We are now ready to test our hypothesis of uniform labelling. To do this, once again the chi-squared test is used.

The expected values (E) are the individual totals of hypothetical grains for each site. The observed values (O) are of course the real grain totals for the same site categories. These must be normalized to the same overall total as the hypothetical grains or vice versa. The individual chi-squared values are totalled and this grand total looked up in a table of chi-squared. The number of degrees of freedom allowed is the number of sites minus the number of sources which in the example in Table 12 would be $8 - 4 = 4$. A further reduction of the degrees of freedom by 1 must be made for each site category in which the hypothetical grain value is less than 5 (see also Section 10.2.1.3).

It can be seen that a square matrix, which would be generated by point-to-point sampling, cannot be tested with validity in this way since there would be no degrees of freedom.

Since it is very unusual for cells and tissues to be uniformly labelled (unless they are dead!) the chi-squared value for the hypothesis of uniform labelling will almost always be very high. This means that there is a low probability that the hypothesis explains the observed grain distribution, i.e. p is almost always less than 0.001.

Table 12. *Idealized cross-fire matrix*

Site → / Source ↓	Cytosol	Nucleus	Granules	Mitochon	Cytosol granules	Cytosol nucleus	Cytosol mitochon	Cytosol granule nucleus	Decays/ unit area
Cytosol	**203**								1
Nucleus	12	**128**				17	18		1
Granules	11		**37**	2	22			12	1
Mitochondria	13			**47**			21		1
Total hypothetical grains (E)	239	128	37	49	22	17	39	12	
Total real grains (0)	187	46	87	21	41	92	50	37	

It is then necessary to change the hypothesis in some way and re-compute chi-squared. Taking Table 12 once again we might for example double all the hypothetical grain values generated from the cytosol and halve those from the nucleus. In other words we now have an hypothesis that the number of decays per unit area from cytosol, nucleus, granules, and mitochondria are 2, 0.5, 1, and 1 respectively. By totalling the revised hypothetical grain numbers for each site and re-testing using chi-squared we can see if our new hypothesis gives a better fit with the real grain numbers.

By now the reader will doubtless have concluded that this manual 'trial and error' approach to hypothesis testing is unthinkably time-consuming and laborious. Fortunately, a computer programme has been written† which contains a chi-squared minimizing sub-routine. This routine sequentially changes the multiples for the hypothetical grain values from each source (i.e. activity values) and re-tests the new site totals, for each change. The programme displays or prints out a 'best grain' matrix and new activity values (plus standard errors) when the lowest chi-squared has been found, that is, when chi-squared has been minimized.

Further description of the data manipulation process and of computation will be given later with reference to an actual set of experimental results.

10.2.2.5. Testing and modifying experimental matrices

Table 13 shows a cross-fire matrix obtained by analysis of EM autoradiographs from an experiment in which rat tail artery tissues had been labelled *in vivo* with the arterial vasodilator drug ^{3}H-hydralazine. It can be seen that there are 8 sources and 16 sites of hypothetical grains. The matrix was made by application of the screen as already described, in other words, the hypothetical grains are distributed in a manner consistant with uniform labelling.

Since there were only 8 hypothetical grains in source 3 (smooth muscle/nucleus) this was combined with source 2 (smooth muscle/mitochondrion) before running the programme. Also, in the interest of avoiding low hypothetical grain (expected) values, the grain totals (hypothetical and real) for certain sites containing similar structures were combined thus:

Site no.: 3 + 16

7 + 10 + 13 + 15

8 + 9

12 + 16.

†A package for hypothetical grain analysis including a 5.25 inch diskette for IBM compatible microcomputers and overlay screens for tritium, iodine-125, and other isotopes can be obtained from TC Associates, Claremont House, 25 Bedford Road, Horsham, West Sussex RH13 5BL, UK.

Table 13. *[³H]-hydralazine-labelled rat tail artery—media and intima 1 h*

Site → / Source ↓	Smooth muscle	Smooth muscle extra-cellular	Elastin	Elastin extra-cellular	Extra-cellular	Elastin/ Smooth muscle	Extra-cellular fibro-blast	Elastin collagen
	(1)	(2)	(3)	(4)	(5)	(6)	(7)	(8)
Smooth muscle cytosol (1)	436	40	0	0	2	2	0	0
Smooth muscle mitochon-drion (2)	29	0	0	0	0	0	0	0
Smooth muscle nucleus (3)	8	0	0	0	0	0	0	0
Marginal collagen (4)	3	3	0	0	2	0	0	0
Marginal elastin (5)	0	0	0	0	0	1	0	0
Collagen (6)	1	1	0	0	5	0	0	0
Elastin (7)	0	1	11	14	3	0	0	0
Extra-cellular (8)	9	25	2	11	50	1	5	0
Total hypothetical grains	486	70	13	25	62	4	5	0
Total real grains	87	26	54	59	2	14	2	1

Hence the matrix which was fitted in the first run was 7×10 (sources × sites).

The results of this computation are shown in the print-out in Fig. 46. A matrix of 'best grains', resulting from the minimizing routine, is shown at the top. The smallest chi-squared obtainable, 27.6, as expected gives $p < 0.001$. The activity values with standard errors are expressed as grains/grid point (remember these were unity before computer fitting). The relative areas are proportional to the number of hypothetical grains recorded for each source. Relative activity (per cent of total) is proportional to grains/grid point × relative area.

Due to the high chi-squared value, an attempt was made to achieve a better fit for the matrix. Referring to the original matrix in Table 13, new combinations were tested:

after intravenous injection

Extra-cellular elastin collagen (9)	Extra-cellular elastin fibroblast (10)	Extra-cellular collagen (11)	Extra-cellular endo-thelium collagen (12)	Fibro-blast (13)	Smooth muscle extra-cellular collagen (14)	Fibro-blast extra-cellular collagen (15)	Endo-thelium lumen (16)	Total
3	0	11	0	0	65	0	0	
0	0	1	0	0	2	0	0	
0	0	0	0	0	0	0	0	
7	0	56	0	0	34	0	0	
4	0	0	0	0	1	0	0	
1	0	203	0	0	4	3	0	
9	0	1	3	0	2	0	0	
11	1	220	1	1	47	7	0	
35	1	492	4	1	155	10	0	1363
93	2	123	0	0	58	2	12	535

Sources: 1 + 2 + 3 (which becomes 'smooth muscle')

Sites: 1 + 2 + 5

3 + 4

6 + 14

7 + 10 + 13 + 15

8 + 9

12 + 16

This 6×7 matrix was fitted and gave the print-out in Fig. 47. It can be seen that there is an excellent fit (chi-squared $= 0.1$; $p > 0.5$). The tritium is seen to be most concentrated (grains/grid point) in the two elastin compartments which between them contain over 48% of total label (relative

Best grains

84	7	0	0	0	0	0	2	0	12
0	0	0	0	0	0	0	0	0	0
1	1	0	0	0	0	3	25	0	15
0	0	12	0	0	0	50	0	0	12
0	0	0	0	1	0	0	45	0	0
0	4	44	56	12	0	36	4	12	8
1	4	0	2	9	2	2	42	0	9

Hypothetical grains

436	40	2	0	2	0	3	11	0	65
37	0	0	0	0	0	0	1	0	2
3	3	0	0	2	0	7	56	0	34
0	0	1	0	0	0	4	0	0	1
1	1	0	0	5	3	1	203	0	4
0	1	11	14	3	0	9	1	3	2
9	25	3	11	50	14	11	220	1	47

RG	87	26	68	59	2	6	93	123	12	58
BG	87	18	58	58	24	3	92	119	12	58
χ^2	0	3	1	0	20	2	0	0	0	0

Total chi-squared 27.6 with 3 degrees of freedom $p < 0.001$

Computer fit with estimated standard errors

Source	Grains/grid point	Relative activity	Relative area
Smooth muscle cytosol	0.194 ± 0.044	20.3 ± 4.1	41.0 ± 0.5
Smooth muscle/mitochon/nucleus	0.002 ± 0.383	0.0 ± 2.6	2.9 ± 0.3
Marginal collagen	0.449 ± 0.377	8.8 ± 6.0	7.7 ± 0.9
Marginal elastin	12.537 ± 1.161	14.1 ± 3.4	0.4 ± 0.1
Collagen	0.226 ± 0.061	9.2 ± 2.2	16.0 ± 0.8
Elastin	4.061 ± 1.290	33.5 ± 0.4	3.2 ± 0.7
Extracellular space	0.193 ± 0.018	14.1 ± 1.6	28.7 ± 0.4

Fig. 46. Computer print-out after 'fitting' the data in Table 13 (and combining certain sources and sites to give a 7×10 matrix as described in Section 10.2.2.5). 'Best grains' is the modified hypothetical grain matrix (using the minimizing subroutine) so that the site totals (BG) agree as closely as possible with those for the real grains (RG) giving the smallest possible chi-squared. It can be seen that in this case the chance that RG and BG arise from the same distribution of activity is less than 0.1%. The programme produces standard errors by generating new sets of hypothetical and real grains in a random manner from Poisson distributions with means equal to the original number of grains. The minimizing procedure is then used to obtain new estimates and the whole process repeated several times

activity). Interestingly, these figures are not very different from the results given for the, apparently, poor fit in Fig. 46. Indeed, it is usually found that decreasing chi-squared to give the conventionally acceptable $p > 0.05$ does not greatly alter the estimates of activity obtained in the previous 'poorer' fit.

Although not used in the example shown, there is a common clue for restructuring matrices which after fitting have not yielded a sufficiently low chi-squared. One merely has to identify which sites individually have large chi-squared components and to combine these which have structures in common. Such combination often reduces the difference between the best grain (BG) and real grain (RG) totals and thus chi-squared.

Best grains

58	0	7	0	0	1	0
2	0	10	0	2	17	0
0	0	24	0	48	0	0
0	0	0	0	0	2	0
16	106	8	0	38	4	12
36	5	21	6	4	96	0

Hypothetical grains

	515	0	69	0	3	12	0
	8	0	34	0	7	56	0
	0	0	2	0	4	0	0
	7	0	4	3	1	203	0
	4	25	2	0	9	1	3
	84	13	48	14	11	220	1
RG	115	113	72	6	94	123	12
BG	114	111	72	6	93	122	13
χ^2	0	0	0	0	0	0	0

Total chi-squared 0.1 with 1 degree of freedom $p > 0.5$

Computer fit with estimated standard errors

Source	Grains/grid point	Relative activity	Relative area
Smooth muscle cell	0.113 ± 0.004	12.7 ± 1.0	43.9 ± 0.2
Marginal collagen	0.317 ± 1.83	6.2 ± 3.6	7.7 ± 0.6
Marginal elastin	12.039 ± 4.106	13.5 ± 5.4	0.4 ± 0.1
Collagen	0.013 ± 0.043	0.5 ± 2.2	16.0 ± 0.7
Elastin	4.247 ± 0.262	34.9 ± 3.0	3.2 ± 0.3
Extracellular space	0.440 ± 0.062	32.1 ± 3.0	28.7 ± 0.2

Fig. 47. Computer print-out after refitting the data in Table 13 (and combining certain sources and sites to give a 6 × 7 matrix as described in Section 10.2.2.5). The 'fit' is now very good with a *p* value greater than 50 per cent. Note that this manipulation, to achieve good fit, has not greatly altered estimates of activity in the two 'elastin' sources which contain the most concentrated radiolabel (compare with Fig. 46)

There is one important eventuality which has not yet been considered, namely the testing of hypotheses that membranes (interfaces which are two-dimensional structures) are labelled. Blackett and Parry (1977) discuss ways in which the overlay screen can be applied to membranes to generate extra grains. (In fact any hypothesis can be tested using this method—it is not necessary to start with an hypothesis of uniform labelling.)

It is important to realize that, when an analysis includes structures which in section are essentially two-dimensional (granules, mitochondria, etc.) and one-dimensional (membranes), the 'relative area' values may not be taken into account for biological interpretation since the dimensions are mixed and thus not comparable. However, 'relative activity' estimates may be compared since they represent the proportion of total label in each compartment whether of square or linear dimensions.

The final result of the method is accepted if biologically plausible. It is tempting to assume that the 'best fit' result is unique and reliable although this is difficult to prove. Nevertheless, there is evidence from two studies that such results are reliable. It has been possible to compare results from identical matrices (unpublished results) analysed by the hypothetical grain method and another method using the same principles of cross-fire measurement but markedly different computation (Downs and Williams 1978). The estimates of activity for each source were closely similar. Another study on indium-111 labelling of platelet granules was able to corroborate closely 'hypothetical grain' predictions of granule labelling using independent biochemical methods (Baker *et al.* 1982).

To the uninitiated reader, the foregoing description doubtless seems somewhat daunting and worse still, very time-consuming. Consider however, the time needed to prepare specimens for EM autoradiography and, in particular, the long exposures which may run into months. There can be no doubt that the extra few hours effort that an analysis of the hypothetical grain type entails can give results of great precision without which the rest of the time invested is hardly justified.

Further reading

Baker, J. R. J., Butler, K. D., Eakins, M. N., Pay, G. F., and White, A. M. (1982). Subcellular localisation of [111]Indium in human and rabbit platelets. *Blood*, **59**, 351–9.

Blackett, N. M. and Parry, D. M. (1973). A new method for analysing electron microscope autoradiographs using hypothetical grain distributions. *J. Cell Biol.* **57**, 9–15.

Blackett, N. M. and Parry, D. M. (1977). A simplified method of 'hypothetical grain' analysis of electron microscope autoradiographs. *J. Histochem. Cytochem.* **25**, 206–14.

Blackett, N. M., Parry, D. M., and Baker, J. R. J. (1980). Isotope decay range distribution curves for use in the analysis of electron microscope autoradiographs. *J. Histochem. Cytochem.* **28**, 1050–4.

Budd, G. C. and Salpeter, M. M. (1969). The distribution of labelled norepinephrine within the sympathetic nerve terminals studied with electron microscopic radio-autography. *J. Cell Biol.* **41**, 21–32.

Downs, A. and Williams, M. A. (1978). An iterative approach to the analysis of E.M. autoradiographs. *J. Microscopy*, **114**, 143–56.

Downs, A. M. and Williams, M. A. (1984). An improved approach to the analysis of autoradiographs containing isolated sources of simple shape: method, theoretical basis and reference data. *J. Microscopy*, **136**, 1–22.

Salpeter, M. M., Bachmann, L., and Salpeter, E. E. (1969). Resolution in electron microscope radioautography. *J. Cell Biol.* **41**, 1–20.

Williams, M. A. (1969). In *Advances in Optical and Electron Microscopy* (ed. R. Barer and V. E. Cosslett) **3**, 219–72. Academic Press, London and New York.

Williams, M. A. (1982). Autoradiography: its methodology at the present time. *J. Microscopy*, **128**, 79–94.

11 Autoradiography in relation to other techniques

Having described the major aspects of autoradiography in some detail, it is prudent to consider the situations in which these techniques may be used to greatest advantage. Autoradiography always involves a period of waiting (for exposure) which can be several hours to many months. The user must therefore see a tangible advantage in the use of autoradiography, as opposed to an alternative colorimetric or other technique.

It is the purpose of this final chapter to identify some of the factors which determine the decision whether to use autoradiography or not.

11.1. Macroautoradiography

In 'grind and find' biochemistry, the products of chromatography or electrophoresis may often be stained with a colour-forming reagent or detected by UV absorption. An example of this is the identification of protein fractions with ninhydrin. Such methods are fast and should be used whenever possible. On the other hand, low molecular weight organic metabolites may not be amenable to these methods and it is then that autoradiography should be used. In this case it is assumed that the starting material is labelled in such a way that all major metabolites contain radionuclide.

Radiometry (i.e. scintillation or gamma counting) of selected organs is commonly used in drug distribution studies. It is usefully employed in conjunction with biochemical analysis of the labelled metabolites. Although such counting permits precise quantification of radiolabel there are many small organs which cannot be excised cleanly, if at all. These, e.g. mouse bile duct, may contain important clues to the routes of metabolism and excretion of the compound studied. If this is so, whole body autoradiography is indispensible.

11.2. LM autoradiography

11.2.1. Fixable labels

Since these chemicals are well preserved *in situ* by aldehyde based fixation, peptide hormones have for many years been recognized as ideal subjects for autoradiography. With the recent explosion in the use of immunocytochemistry, using peroxidase as chromogen or gold/silver labels, it seems logical or indeed probable that autoradiography will largely be superseded in this area. It should be understood, however, that the apparent advantages of high speed and high resolution attached to immunocytochemistry may well be offset initially by time spent trying to get the technique to work.

11.2.2. Diffusible substances

Most cytochemical methods rely upon aqueous treatment of the specimen. Even those enzyme cytochemical methods in which substrate and coenzyme are trapped in a gel must allow some translocation of molecules in aqueous phase.

The LM diffusible compound autoradiographic methods are liquid-free up to the stage of photographic development. They must therefore be considered to offer the best fidelity for localization of diffusibles albeit for exogenous substances only.

11.2.3. Receptor autoradiography

There is probably no real alternative to autoradiography for receptor localization. However, the widespread use of ^{125}I labelled ligands with scant regard for the effect of iodination on biological activity in many cases is a cause for concern.

11.3. EM autoradiography

EM immunocytochemistry in its many forms may be used for the ultrastructural localization of proteins and peptides of exogenous as well as endogenous origin. Although immuno-gold deposits permit superior resolution to autoradiographic silver grains, there has been very little attempt so far to quantify EM immunocytochemistry. Where autoradiography may allow a range of metabolites to be visualized (if they are fixed) immunocytochemistry reveals only fully antigenic epitopes. Hence the objectives of the experiment will determine the most suitable method.

For inorganic substances, X-ray microanalysis may localize elements of atomic number 11 (i.e. Na) and above. Signals from 'organic elements' (i.e. H, C, N, O) in a compound of interest are lost in the biological organic matrix. This is equally true of electron energy loss spectroscopy (although in non-biological specimens this is capable of detecting C, O, and N). In addition, the threshold concentrations for analytical detection of many elements in ultra-thin sections of resin embedded fixed tissues are rarely reached. This is also often the case for ultra-thin unfixed cryosections.

Although in Chapter 8 EM autoradiography was shown to be a relatively laborious technique in terms of time needed to get an image, it should be remembered that in most autoradiographic conditions a single atom of radionuclide has the potential to produce a latent image which may be made visible. No other method of cytochemical detection can compete with this level of ultimate sensitivity.

.

Index

Milton Keynes UK
Ingram Content Group UK Ltd.
UKHW040051071024
449327UK00019B/463